ELEMENTARY MATHEMATICS

Grade 6

Manpekin

authorHOUSE

AuthorHouse™
1663 Liberty Drive
Bloomington, IN 47403
www.authorhouse.com
Phone: 1 (800) 839-8640

Published by AuthorHouse 04/06/2020

ISBN: 978-1-7283-4634-2 (sc)
ISBN: 978-1-7283-4633-5 (e)

Library of Congress Control Number: 2020903692

Print information available on the last page.

Any people depicted in stock imagery provided by Getty Images are models, and such images are being used for illustrative purposes only.
Certain stock imagery © Getty Images.

This book is printed on acid-free paper.

CONTENTS

UNIT 1
SETS

Learning points

In this unit, we shall

- learn about sets;
- learn about Venn diagrams;
- learn about rational and irrational numbers; and
- discuss sets of prime numbers and points.

LESSON 1.1

DEFINITION AND MEMBERSHIP OF SETS

A set is a collection of objects or things. For example, a set of days of the week contains 7 objects: Sunday, Monday, Tuesday, Wednesday, Thursday, Friday, and Saturday.

Braces are used to show a set. For example, set A = {a, b, c}.

A set can have members. For example, if set A = {1, 2, 3}, then the members of set A are 1, 2, and 3. The members of a set are also called the elements of a set.

The order in which the elements of a set are written does not matter. For example, set A above can also be written as set A = {3, 1, 2}, or set A = {2, 3, 1}.

Symbols $\in$ and $\notin$ are used to show membership of sets.

To indicate that an element is a member of a set, use the symbol $\in$.

To indicate that an element is *not* a member of a set, use the symbol $\notin$.

Let set B = {4, 5, 6} and set D = {7, 8, 9}.

These symbols can be used to write that 4 is a member of set B as $4 \in B$ and 8 is a member of set D as $8 \in D$.

They can also be used to write that 7 is not a member of set B as $7 \notin B$ and 6 is not a member of set D as $6 \notin D$.

Sample Work 1

Use $\in$ or $\notin$ to get a true statement: 1___ {5, 4, 1}.

Solution

  1 _$\in$_ {5, 4, 1} since 1 is a member of {5, 4, 1}.

Sample Work 2

Use ∈ or ∉ to get a true statement: 9 ___ {6, 7, 8}.

Solution

9 _∉_ {6, 7, 8} since 9 is not a member of {6, 7, 8}.

Practice

1. List the set of months of the year.

2. List the set of the last 4 months of the year.

3. List the set of the first 3 days of the week.

Use ∈ or ∉ to get a true statement.

4. 7___ {1, 2, 3, 7}

5. 3 ___ {1, 4, 5,}

6. 5 ___ {5, 6, 7, 8}

7. 6 ___ {7, 8, 9}

8. 6 ___ {6, 7, 8}

9. 8 ___ {10, 8, 1}

10. 9 ___ {1, 9, 10}

11. 8 ___ {2, 3, 4}

Tell whether the following statements are true or false.

12. 9 ∈ {4, 9, 7, 3, 6}

13. {7, 5} ∈ {10, 8, 7, 4, 5, 2}

14. 5 ∈ {6, 7, 8}

15. 4 ∈ {3, 5, 7, 8}

16. $7 \notin \{8, 9, 10, 11\}$

17. $13 \notin \{10, 11, 12, 13\}$

18. $9 \notin \{2, 3, 4, 5\}$

19. $8 \notin \{6, 7, 9\}$

20. $12 \notin \{10, 11, 12\}$

21. $1 \notin \{2, 3, 4, 5\}$

22. $1 \in \{4, 3, 2, 1\}$

23. $9 \in \{6, 7, 8, 9\}$

24. $8 \in \{6, 7, 8\}$

25. $4 \in \{1, 2, 3, 4\}$

LESSON 1.2

EQUAL SETS AND EQUIVALENT SETS

Two sets are equal when they have the same equal elements.

For example, let set A = {5, 6, 7} and set B = {5, 6, 7}. Set A and set B are equal since they both have the same equal elements.

Sample Work 1

Are sets E and F equal?

E = {3, 4, 5}

F = {3, 4, 5}

Solution

Set E and set F are equal since they have the same equal elements.

Sample Work 2

Set A = {2, 5, 9} and set B = {5, 2, 9}. Show whether A is equal to B.

Solution

A = B. They both have the same equal elements.

Notice that the order in which the elements of equal sets are written does not matter.

Two sets are equivalent when they have the same number of elements.

For example, if set C = {4, 5, 6} and set D = {7, 8, 9}, then set C and set D are equivalent. Both sets have 3 members each.

You may use the symbol ↔ to show equivalent sets.

Sample Work 3

Are sets G and H equivalent?

$G = \{9, 3, 1\}\ H = \{4, 6, 5\}$

Solution

Set G is equivalent to set H since they both have the same number of elements.

Sample Work 4

Set $P = \{1, 6, 7\}$ and set $Q = \{1, 8, 7\}$. Show whether P is equivalent to Q.

Solution

$P \leftrightarrow Q$ since they both have the same number of elements. They have 3 elements each.

Sample Work 5

$D = \{a, b, c\}$ and $E = \{a, b, c\}$. Show that D is equivalent to E.

Solution

$D \leftrightarrow E$ since they both have the same number of elements.

Practice

State or show that the sets are equal.

1. $A = \{1, 2, 3, 4\}\ B = \{1, 2, 3, 5\}$

2. $C = \{3, 4, 5, 6\}\ D = \{3, 1, 4, 6\}$

3. $E = \{2, 3, 4, 5\}\ F = \{2, 4, 5, 3\}$

4. $G = \{5, 6, 7\}\ H = \{7, 5, 6\}$

5. $I = \{10, 11, 12\}\ J = \{12, 9, 10\}$

6.   K = {1, 2, 3} L = {1, 2, 3}

7.   M = {2, 3, 5} N = {5, 2, 3}

8.   O = {3, 4, 5} P = {5, 3, 4}

9.   Q = {5, 6, 8} R = {5, 6, 8, 9}

10. S = {7, 8, 9, 1} T = {1, 7, 8, 9}

11. U = {5, 4, 8, 9} V = {9, 4, 8, 5}

12. W = {10, 11, 12} X = {12, 13, 14}

State or show that the sets are equivalent.

1.   A = {1, 2, 3, 4} B = {1, 2, 3, 5}

2.   C = {3, 4, 5, 6} D = {3, 1, 4, 6}

3.   E = {2, 3, 4, 5} F = {2, 4, 5, 3}

4.   G = {5, 6, 7} H = {7, 5, 6}

5.   I = {10, 11, 12} J = {12, 9, 10}

6.   K = {1, 2, 3} L = {1, 2, 3}

7.   M = {2, 3, 5} N = {5, 2, 3}

8.   O = {3, 4, 5} P = {5, 3, 4}

9.   Q = {5, 6, 8} R = {5, 6, 8, 9}

10. S = {7, 8, 9, 1} T = {1, 7, 8, 9}

11. U = {5, 4, 8, 9} V = {9, 4, 8, 5}

12. W = {10, 11, 12} X = {12, 13, 14}

LESSON 1.3

INTERSECTION AND UNION OF SETS

The intersection of 2 sets is the set of all members that are common to both sets. The symbol ∩ is used to show intersection of sets.

Sample Work 1

Find the intersection of the sets.

$A = \{1, 2, 3\}$ $B = \{1, 4, 3\}$

Solution

Look at the two sets and identify the elements of set A and set B that are common to each other.

$A = \{1, 2, 3\}$ $B = \{1, 4, 3\}$

The elements 1 and 3 are common to both set A and set B.

So $A \cap B = \{1, 3\}$.

Sample Work 2

Find the intersection.

$E = \{6, 7, 8, 9\}$ $F = \{6, 4, 8, 9\}$

Solution

Look at the two sets and identify the elements of set E and set F that are common to each other.

$E = \{6, 7, 8, 9\}$ $F = \{6, 4, 8, 9\}$

The elements, 6, 8, and 9 are members of both set E and set F.

So $E \cap F = \{6, 8, 9\}$.

Sample Work 3

What is the intersection of R = {x is a set of odd numbers between 1 and 7}, and S = {x is whole numbers between 0 and 4}?

Solution

Odd numbers between 1 and 7: 3, 5

Whole numbers between 0 and 4: 1, 2, 3

R = {3, 5}, and S = {1, 2, 3}

So, R $\cap$ S = {3}.

The union of two sets is the set of all members that are in either set—that is, when all the elements of the two sets are combined.

The symbol U is used to show union of sets.

Sample Work 4

Find the union of the sets.

$T = \{7, 8, 9\}$ $V = \{4, 7, 8\}$

Solution

$T = \{7, 8, 9\}$ $V = \{4, 7, 8\}$

The set of all the elements of the two sets combined are {4, 7, 8, 9}.

So $T \cup V = \{4, 7, 8, 9\}$.

Sample Work 5

Find the union of the sets.

$G = \{1, 3, 4\}$ $H = \{1, 3, 5\}$

Solution

$G = \{1, 3, 4\}$ $H = \{1, 3, 5\}$

The elements 1, 3, 4, and 5 are in either set G or set H.

So $G \cup H = \{1, 3, 4, 5\}$.

Sample Work 6

What is the union of A = {x is a set of counting numbers between 3 and 8} and B = {2, 4, 6, 8}?

Solution

A set of counting number between 3 and 8: {4, 5, 6, 7}.

$A = \{4, 5, 6, 7\}$ $B = \{2, 4, 6, 8\}$.

Since "union of a set" means "every member that is in either set," the union is every member from A plus every member from B.

So $A \cup B = \{2, 4, 5, 6, 7, 8\}$.

Practice

Find the intersection and union of the sets.

1. $A = \{3, 5, 7\}$ $B = \{7, 3, 6\}$

2. $S = \{1, 2, 3, 5\}$ $T = \{2, 3, 5, 6\}$

3. $Y = \{5, 6, 7, 8\}$ $Z = \{5, 7, 8, 9\}$

4. $C = \{7, 9, 6, 1\}$ $D = \{6, 7, 1\}$

5. $E = \{5, 6, 3\}$ $F = \{5, 6, 1\}$

6. $G = \{8, 2, 4, 5, 7\}$ $H = \{9, 4, 5, 6, 1\}$

7. $I = \{9, 10, 11, 12\}$ $J = \{10, 12, 13, 14\}$

8. $K = \{1, 2, 3, 4\}$ $L = \{3, 4, 6, 8\}$

9. $M = \{6, 7, 8, 9\}$ $N = \{5, 6, 7, 8\}$

10.　$P = \{10, 11, 12, 13\}$　$Q = \{12, 13, 14, 15\}$

11.　What is the union if set A = {x is a set of counting numbers from 2 and 5 inclusive} and set B = {x is a set of odd numbers between 3 and 9}?

12.　Find the union of A = {x is a set of whole numbers between 9 and 14} and B = {x is a set of single-digit positive numbers between 7 and 11}.

13.　What is the intersection if set A = {x is a set of counting numbers less than 7} and set B = {x is a set of counting numbers less than 5}?

14.　Find the intersection of set A = {x is a set of even numbers between 5 and 9} and B = {x is a set of even numbers between 1 and 6}.

15.　What is the intersection if set A = {x is whole numbers less than 6} and set B = {x is whole numbers less than 4}?

LESSON 1.4

REPLACEMENT SET AND SOLUTION SET IN SET NOTATION

Replacement set is the set of values that can be used to replace a variable. Solution set is a number chosen from the replacement set that satisfies a given inequality. To find the solution set, substitute each value from the replacement set and solve the inequality.

Sample Work 1

Find the solution set: x < 10, if the replacement set is the set of counting numbers less than 10.

Solution

Inequality: x < 10.

The set of counting numbers up to 10: {1, 2, 3, 4, 5, 6, 7, 8, 9, 10}

The set of counting numbers less than 10: {1, 2, 3, 4, 5, 6, 7, 8, 9}

The solution set: {1, 2, 3, 4, 5, 6, 7, 8, 9}.

Sample Work 2

Find the solution set: x > 5, if the replacement set is the set of whole numbers greater than 5 and less than 12.

Solution

The set of whole numbers from 5 to 12: {5, 6, 7, 8, 9, 10, 11, 12}

The set of whole numbers greater than 5 and less than 12: {6, 7, 8, 9, 10, 11}

So, the solution set = {6, 7, 8, 9, 10, 11}.

Sample Work 3

Find the solution set for x + 3 < 15 from the replacement set {4, 8, 11, 16}.

Solution

Inequality: x + 3 < 15; Replacement set: {4, 8, 11, 16}

Substitute the values {4, 8, 11, 16} into the equation x + 3 < 15.

When x = 4: 4 + 3 = 7

When x = 8: 8 + 3 = 11

When x = 11: 11 + 3 = 14

When x = 16: 16 + 3 = 19

Plug the values into the equation and pick out the true statement.

4 + 3 = 7 < 15: true

8 + 3 = 11 < 15: true

11 + 3 = 14 < 15: true

16 + 3 = 19 < 15: false

Thus, the solution set = {4, 8, 11}.

Sample Work 4

Find the solution set for x + 5 > 20 from the replacement set {5, 10, 15, 20}.

Solution

Equation: x + 5 > 20; replacement set: {5, 10, 15, 20}.

Substitute the values {5, 10, 15, 20} into the equation x + 5 > 20

When x = 5: 5 + 5 = 10

When x = 10: 10 + 5 = 15

When x = 15: 15 + 5 = 20

When x = 20: 20 + 5 = 25

Plug in the values into the equation and pick up the true statement.

5 + 5 = 10 > 20: false

10 + 5 = 15 > 20: false

15 + 5 = 20 > 20: false

20 + 5 = 25 > 20: true

Thus, the solution set = {20}.

Practice

1. Find solution set of x < 17 when the replacement set is {4, 6, 8, 10}.

2. If the replacement set is the set of whole numbers greater than 1 and less than 6, find the solution set of x < 4.

3. If the replacement set is the set of whole numbers greater than 5 and less than 10, find the solution set of x < 9.

4. If the replacement set is the set of whole numbers less than 8, find the solution set of x > 2.

5. If the replacement set is the set of even numbers between 1 and 15, find the solution set of x > 10.

6. If the replacement set is the set of even numbers between 15 and 20, find the solution set of x < 19.

7. If the replacement set is the set of odd numbers between 20 and 36, find the solution set of x < 35.

8. If the replacement set is the set of counting numbers less than 14, find the solution set of x > 7.

LESSON 1.5

VENN DIAGRAM

A Venn diagram is used to show set relationship.

Intersection of the sets

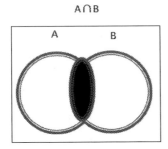

A∩B

The shaded or overlapped portion is the intersection of sets A and B.

Union of the sets

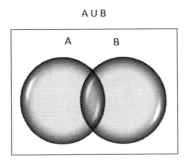

A∪B

The shaded portion is the union of sets A and B

Sample Work 1

Use the Venn diagram to write the members of V, U, U ∩ V, U ∪ V.

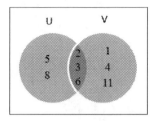

solution

V = {1, 2, 3, 4, 6, 11} U = {2, 3, 5, 6, 8}

U ∩ V = {2, 3, 6}

U ∪ V = {1, 2, 3, 4, 5, 6, 8, 11}

Practice

1. Construct a Venn diagram using set *E* = {4, 5, 6, 8}, set *F* = {3, 1, 5, 6}.

2. Use the Venn diagram to write the members of P, Q, P ∩ Q, P ∪ Q.

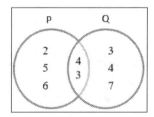

3. Use the Venn diagram to write the elements of T, S, S ∩ T, S ∪ T.

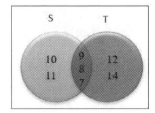

4. Use the Venn diagram to write the members of G, H, G ∩ H, G ∪ H.

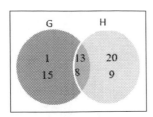

5. Construct a Venn diagram using set A = {4, 5, 6, 8}, set B = {3, 1, 5, 6}, A ∩ B = {5, 6}

6. Using a Venn diagram, list the elements of C ∩ D, C ∪ D.

Set C = {1, 4, 6, 7}

Set D = {2, 3, 4, 7}

LESSON 1.6

RATIONAL AND IRRATIONAL NUMBERS

A rational number is a number that can be written as a ratio. It can also be written as a fraction in which both the numerator (the number on top) and the denominator (the number on the bottom) are whole numbers. For example, the number 8 is a rational number since it can be written as the fraction, $\frac{8}{1}$.

Likewise, 0.75 is a rational number since it can be written as a fraction, $\frac{3}{4}$.

Every whole number is a rational number, because any whole number can be written as a fraction. For example, 4 can be written as $\frac{4}{1}$, and 63 can be written as $\frac{63}{1}$.

Sample Work 1

Show that $6 - \sqrt{4}$ is a rational number.

Solution

$6 - \sqrt{4} = 6 - 2 \; \underline{\sqrt{4} = 2}$ (simplifying the square root to 2) $= 4$

4 can be written as a quotient of 4 and 1 as $\frac{4}{1}$.

So, $6 - \sqrt{4}$ is a rational number.

Sample Work 2

Show that $15 - \sqrt{9}$ is a rational number.

Solution

$15 - \sqrt{9} = 15 - 3 \; \underline{\sqrt{9} = 3}$ (simplifying the square root to 3) $= 12$

12 can be written as a quotient of 12 and 1 as $\frac{12}{1}$.

So 15 - $\sqrt{9}$ is rational number.

Sample Work 3

Show that 22 - $\sqrt{16}$ is a rational number.

Solution

22 - $\sqrt{16}$ = 22 − 4 $\underline{\sqrt{16} = 4}$ (simplifying the square root to 4) = 18

18 can be written as a quotient of 18 and 1 as $\frac{18}{1}$.

So 22 - $\sqrt{16}$ is rational.

Sample Work 4

Is the number $\sqrt{25}$ rational?

Yes, since $\sqrt{25}$ can be simplified to the whole number, which can be written as $\frac{5}{1}$, a quotient of two whole numbers.

Sample Work 5

Number	As a Fraction	Rational?
$\sqrt{7}$	(square root of 7)?	no
5	$\frac{5}{1}$	Yes
$\sqrt{13}$	(square root of 13)?	no
0.46	$\frac{6}{13}$	yes
.01	$\frac{1}{100}$	Yes
$\sqrt{26}$	(square root of 26)?	no

Practice

1.   Show that $7 + \sqrt{25}$ is a rational number.

2.   Show that $21 + \sqrt{9}$ is a rational number.

3.   Show that $\sqrt{49} + 12$ is a rational number.

4.   Show that $11 - \sqrt{81}$ is a rational number.

5.   Show that $100 - \sqrt{64}$ is a rational number.

6.   Show that $30 - \sqrt{16}$ is a rational number.

7.   Show that $3 - \sqrt{15}$ is a rational number.

8.   Show that $7 - \sqrt{15}$ is a rational number.

9.   Show that $12 - \sqrt{26}$ is a rational number.

10. Show that $\sqrt{16} - 2$ is a rational number.

11.

Number	As a Fraction	Rational?
7	$\dfrac{7}{1}$	
1.75	$\dfrac{7}{4}$	
.001	$\dfrac{1}{1000}$	
$\sqrt{3}$	(square root of 3)?	
$\sqrt{4}$	(square root of 4)?	

Irrational Numbers

An irrational number is a number that is not considered rational. An irrational number cannot be expressed as a fraction but can be written as a decimal. It has endless non-repeating digits to the right of the decimal.

In other words, an irrational number cannot be written as a fraction with whole numbers in the numerator and denominator.

For example, $\sqrt{2}$ = 1.41 and π = 3.14 are irrational numbers.

Irrational numbers can be clearly seen/indicated on the number line where there are infinite numbers of irrational numbers. For example, between 0 and 1.

Sample Work 1

Unlike $9-\sqrt{9}$, you cannot simplify $5-\sqrt{5}$.

Sample Work 2

Unlike $9-\sqrt{9}$, you cannot simplify $7-\sqrt{7}$.

Sample Work 3

Is the number $\sqrt{35}$ irrational?

Yes, because $\sqrt{35}$ cannot be simplified to a whole number, which means it cannot be expressed as a quotient of two whole numbers.

Sample Work 4

Is the number $\sqrt{15}$ irrational?

Yes, because $\sqrt{15}$ cannot be simplified to a whole number, which means it cannot be expressed as a quotient of two whole numbers.

Practice

Tell or show whether the expression is irrational

1. Is the number $\sqrt{17}$ irrational?

2. Is the number $\sqrt{20}$ irrational?

3. Is the number $\sqrt{81}$ irrational?

4. Is the number $\dfrac{\sqrt{4}}{5}$ irrational?

5. Is the number $\dfrac{\sqrt{5}}{6}$ irrational?

6. Is the number $\dfrac{\sqrt{9}}{7}$ irrational?

7. Is the number $\dfrac{\sqrt{49}}{7}$ irrational?

8. Is the number $\dfrac{\sqrt{21}}{6}$ irrational?

9. Is the number $\dfrac{\sqrt{5}}{2}$ irrational?

10. Is the number $\dfrac{\sqrt{18}}{3}$ irrational?

SETS OF PRIME NUMBERS AND POINTS

A whole number that is greater than 1 and whose factors are only 1 and itself is called a *prime number*. Whole numbers are counting numbers, including 0, 1, 2, 3, 4, 5 … 54, 55, 56, 57, 58, 59, 60, 61, 62, 63, 64, 65, 66, 67, 68, 69, 70, 71, 72, 73, 74, 75, 76, 77, 78, 79, 80 …

A prime number has no lower multiples than 1. For example, 17 is a prime number since its factors are only 1 and 17.

The set of numbers such as {2, 3, 5, 7, 11, 13, 17, 19} is a set of prime numbers from 2 to 19 inclusive.

Sample Work 1

Find the set of prime numbers between 70 and 80.

Solution

Set of whole numbers between 70 and 80: {71, 72, 73, 74, 75, 76, 77, 78, 79}.

There are 9 whole numbers between 70 and 80.

Identify and select the prime numbers: 71, 73, 75, 77, 79 (ignoring the even numbers).

There are 5 prime numbers between 70 and 80: 71, 73, 75, 77, 79.

The set of all prime numbers between 70 and 80: {71, 73, 75, 77, 79}.

Sample Work 2

What is the least prime number between 38 and 51?

Solution

The set of whole numbers between 38 and 51: {39, 40, 41, 42, 43, 44, 45, 46, 47, 48, 49, 50}.

The set of prime numbers between 38 and 51: {41, 43, 47, 49}.

The least prime number between 38 and 51: {41}.

Sample Work 3

What is the greatest prime number between 4 and 15?

Solution

The set of whole numbers between 4 and 15: {5, 6, 7, 8, 9, 10, 11, 12, 13, 14}.

The set of prime numbers between 4 and 15: {5, 7, 11, 13}.

The greatest prime number between 4 and 15: {13}.

Practice

1. List all prime numbers between 30 and 60.

2. List all prime numbers between 15 and 30.

3. List all prime numbers between 10 and 25.

4. List all prime numbers between 3 and 7.

5. List all prime numbers between 50 and 80.

6. What is the greatest prime number between 12 and 20?

7. What is the greatest prime number between 10 and 19?

8. What is the greatest prime number between 1 and 7?

9. What is the greatest prime number between 35 and 50?

10. What is the greatest prime number between 70 and 90?

11. What is the least prime number between 1 and 10?

12. What is the least prime number between 12 and 25?

13. What is the least prime number between 6 and 10?

14. List all prime numbers from 1 to 100.

Sets of Points

A set is a collection of things having common property. At least 2 points can make a line. A line is a set of points. A line is a collection of multiple points. A circle is a set of points.

UNIT 2

NUMBER BASES

Learning Points

In this unit, we shall

- learn how to change from one base to another; and
- learn how to add, subtract, and multiply in bases 2, 3, and 5.

CHANGING BASE 10 TO BASES 2 AND 5

To change from a larger base to a smaller base, divide.

For example, to change from base 10 to base 5, divide the number by 5.

Sample Work 1

Change 53_{10} to base 2.

Solution

$53_{10} =$ —— base 2

Divide 53 by 2 successively until the remainder is less than 2.

2	53	1
2	26	0
2	13	1
2	6	0
2	3	1
	1	

← Dividing by 2 until the remainder is less than 2.

$53_{10} = 110101_2$

Sample Work 2

Change 67_{10} to base 2.

Solution

Divide $67 \div 2$ successively until the remainder is less than 2.

2	67	1
2	33	1
2	16	0
2	8	0
2	4	0
2	2	0
	1	

$67_{10} = 1000011_2$

Sample Work 3

Change 94_{10} to base 5.

Solution

Divide 94 by 5 successively until the remainder is less than 5.

5	94	4
5	18	3
	3	

$94_{10} = 334_5$

Practice

Change the following base ten numerals to base 2.

1. 91_{10}

2. 84_{10}

3. 76_{10}

4. 66_{10}

5. 54_{10}

6. 45_{10}

7. 25_{10}

8. 39_{10}

9. 49_{10}

10. 55_{10}

11. 32_{10}

12. 37_{10}

13. 18_{10}

14. 20_{10}

15. 51_{10}

16. 19_{10}

17. 92_{10}

18. 62_{10}

19. 77_{10}

20. 68_{10}

Change to base 5:

1. 96_{10}

2. 87_{10}

3. 77_{10}

4. 64_{10}

5. 52_{10}

6. 44_{10}

7. 32_{10}

8. 22_{10}

9. 10001_{10}

10. 101_{10}

11. 21_{10}

12. 81_{10}

13. 95_{10}

14. 39_{10}

15. 28_{10}

LESSON 2.2

CHANGING BASE 5 TO BASE 10

To change from a smaller base to a larger base, multiply.

Sample Work 1

Change 2012_5 to base 10.

Solution

2012_5

$2012_5 = (2 \times 5^3) + (0 \times 5^2) + (1 \times 5^1) + (2 \times 5^0)$

$= (2 \times 125) + (0 \times 25) + (1 \times 5) + (2 \times 1)$

$= 250 + 0 + 5 + 2$

$= 250 + 5 + 2$

$= 250 + 7$

$= 257_{10}$

Sample Work 2

Change 341_5 to base 10.

Solution

341_5

$341_5 = (3 \times 5^2) + (4 \times 5^1) + (1 \times 5^0)$

$$= (3 \times 25) + (4 \times 5) + (1 \times 1)$$

$$= 75 + 20 + 1$$

$$= 75 + 21$$

$$= 96_{10}$$

Sample Work 3

Convert to base 10: 432_5

Solution

Convert to base 10: 432_5

$$432_5 = 4 \times 5^2 + 3 \times 5^1 + 2 \times 5^0$$

$$= 4 \times 25 + 3 \times 5 + 2 \times 1$$

$$= 100 + 15 + 2$$

$$= 117_{10}.$$

Sample Work 4

Convert to base 10: 403_5.

Solution

Convert to base 10: 403_5

$$403_5 = 4 \times 5^2 + 0 \times 5^1 + 3 \times 5^0$$

$$= 4 \times 25 + 0 \times 5 + 3 \times 1$$

$$= 100 + 0 + 3$$

$$= 103_{10}.$$

Practice

Change to base 10.

1. 40_5

2. 34_5

3. 400_5

4. 311_5

5. 401_5

6. 113_5

7. 2011_5

8. 114_5

9. 422_5

10. 224_5

11. 333_5

12. 444_5

13. 330_5

14. 104_5

15. 14_5

16. 312_5

17. 303_5

18. 402_5

19. 422_5

20. 2222_5

21. 11012_5

22. 3114_5

23. 10043_5

24. 1114_5

25. 2304_5

LESSON 2.3

ADDING AND SUBTRACTING IN BASES 2 AND 3

Adding and subtracting in bases 2 and 3 is like adding and subtracting in whole numbers.

Sample Work 1

Add $\quad 1011_2$

$\quad\quad + 101_2$

Solution

$\quad\quad 1011_2 \longleftarrow$ Adding in 2s

$\quad\quad + 101_2$

$\quad\quad 10000_2$

Sample Work 2

Add $111_2 + 101_2$

Solution

$111_2 + 101_2$

$\quad\quad 111_2$

$\quad + 101_2 \longleftarrow$ Adding in 2s

$\quad\quad 1100_2$

Sample Work 3

Add $\quad 221_3$

$\quad\quad + 22_3$

Solution

$$221_3$$
$$+22_3 \longleftarrow \text{Adding in 3s}$$
$$\overline{1020_3}$$

Sample Work 4

Subtract $1010_2 - 101_2$

Solution

$1010_2 - 101_2$

$$1010_2$$
$$- 101_2 \qquad \text{1 borrowed is equal to 2.}$$
$$\overline{101_2}$$

Sample Work 5

Subtract $\qquad 212_3$
$$\overline{-122_3}$$

Solution

$$212_3$$
$$-122_3 \qquad \text{1 borrowed is equal to 3}$$
$$\overline{20_3}$$

Practice

Add or subtract as shown.

1. 101_2
   $+11_2$

2. 1110_2
   $+111_2$

3. 1011_2
   $+111_2$

4. 1001_2
   $+111_2$

5. $110_2 - 11_2$

6. $1001_2 - 111_2$

7. $1011_2 - 111_2$

8. $211_3 + 21_3$

9. $2011_3 + 211_3$

10. $1212_3 + 112_3$

11. $1212_3 + 212_3$

12. $2020_3 + 202_3$

13. $2120_3 - 2110_3$

14. $112_3 - 22_3$

15. $121_3 - 22_3$

16. $221_3 - 122_3$

LESSON 2.4

ADDING AND SUBTRACTING IN BASE 5

We shall add and subtract in 5s.

Sample Work 1

Add $432_5 + 143_5$

Solution

$432_5 + 143_5$

$$
\begin{array}{r}
432_5 \\
+\ 143_5 \\
\hline
1130_5
\end{array}
\qquad \text{Adding in 5s}
$$

Sample Work 2

Add $33_5 + 12_5$

Solution

$33_5 + 12_5$

$$
\begin{array}{r}
33_5 \\
+\ 12_5 \\
\hline
100_5
\end{array}
$$

Sample Work 3

Add $24_5 + 12_5$

Solution

$24_5 + 12_5$

$$
\begin{array}{r}
24_5 \\
+\ 12_5 \\
\hline
41_5
\end{array}
$$

Sample Work 4

Subtract $423_5 - 244_5$

Solution

$423_5 - 244_5$

$$
\begin{array}{r}
423_5 \\
-244_5 \\
\hline
124_5
\end{array}
$$

Sample Work 5

Subtract $32_5 - 14_5$

Solution

$32_5 - 14_5$

$$
\begin{array}{r}
32_5 \\
-14_5 \\
\hline
13_5
\end{array}
$$

Practice

Add or subtract in base 5 as shown.

1. $43_5 - 4_5$

2. $341_5 + 234_5$

3. $41_5 + 4_5$

4. $143_5 + 443_5$

5. $234_5 + 234_5$

6. $323_5 + 431_5$

7. $2144_5 + 344_5$

8. $344_5 - 133_5$

9. $422_5 — 234_5$

10. $\begin{aligned} 3111_5 \\ \underline{—223_5} \end{aligned}$

11. $\begin{aligned} 323_5 \\ \underline{— 44_5} \end{aligned}$

12. $\begin{aligned} 3211_5 \\ \underline{— 432_5} \end{aligned}$

13. $\begin{aligned} 444_5 \\ \underline{+ 131_5} \end{aligned}$

14. $\begin{aligned} 333_5 \\ \underline{+ 22_5} \end{aligned}$

15. $\begin{aligned} 322_5 \\ \underline{—144_5} \end{aligned}$

16. $\begin{aligned} 34_5 \\ \underline{+ 24_5} \end{aligned}$

17. $\begin{aligned} 31_5 \\ \underline{—14_5} \end{aligned}$

18. $\begin{aligned} 12_5 \\ \underline{—3_5} \end{aligned}$

19. $\begin{aligned} 112_5 \\ \underline{— 23_5} \end{aligned}$

20. $\begin{aligned} 213_5 \\ \underline{—132_5} \end{aligned}$

LESSON 2.5

MULTIPLYING NUMBERS IN BASE 5

To multiply in base 5, start by multiplying as you would in base 10.

Sample Work 1

Multiply 314_5
           $\text{x } 3_5$

Solution

314_5
$\text{x } 3_5$

Start by multiplying as you would in base 10. So 3 times 4 is 12, and 12 is 22_5. Write 2 and carry 2.

   2
314_5
  $\text{x } 3_5$
   2

Next, multiply 3 times 1 to get 3 and add the 2 you carried using base 5 addition to get 5. Five in base 5 is 10_5. So write 0 and carry 1.

   1
314_5
  $\text{x } 3_5$
  02

Finally, multiply 3 times 3 to get 9 and add the 1 you carried using base 5 addition to get 10. Ten in base 5 is 20_5. Write the 20 as thousand and hundred digits in the product of the multiplication.

$$\begin{array}{r} 314_5 \\ \times\ 3_5 \\ \hline 2002_5 \end{array}$$

Sample Work 2

Multiply.

$$\begin{array}{r} 43_5 \\ \times 4_5 \\ \hline \end{array}$$

Solution

$$\begin{array}{r} 43_5 \\ \times\ 4_5 \\ \hline \end{array}$$

$$\begin{array}{r} 2 \\ 43_5 \\ \times\ 4_5 \\ \hline 2 \end{array}$$

4 x 3 = 12 ÷5 = 22_5, write down the 2 and carry 2

Next, multiply 4 x 4 = 16 + 2 you carried = 18. Number 18 in base 5 is 33 base 5. Write down the 33 as hundred and ten digits in the product of the multiplication. Use base 5 addition to write the answer.

$$\begin{array}{r} 2 \\ 43_5 \\ \times\ 4_5 \\ \hline 332_5 \end{array}$$

Practice

Multiply

1.
$$\begin{array}{r} 21_5 \\ \times 11_5 \\ \hline \end{array}$$

2. 43_5
     $\times 14_5$
     _____

3. 113_5
      $\times 4_5$
     _____

4. 41_5
      $\times 3_5$
     _____

5. 40_5
      $\times 3_5$
     _____

6. 13_5
      $\times 4_5$
     _____

7. 423_5
      $\times 2_5$
     _____

8. 34_5
     $\times 21_5$
     _____

9. 222_5
      $\times 3_5$
     _____

10. 44_5
      $\times 2_5$
     _____

11. 23_5
      $\times 4_5$
     _____

12. 24_5
      $\times 3_5$
     _____

13. 34_5
      $\times 3_5$
     _____

UNIT 3

OPERATIONS ON DECIMALS

Learning points

In this unit, we shall

- learn how to add, subtract, multiply, and divide decimal numbers;
- learn how to round off decimals;
- learn how to change fractions to decimals; and
- learn how to estimate decimals and whole numbers.

LESSON 3.1

ADDING AND SUBTRACTING DECIMALS

To add or subtract decimal numbers, line up the decimal points to add or subtract. They can be written in either column (vertical) or row (horizontal) form. Column form is the vertical arrangement of the problem while row form is the horizontal arrangement of the problem.

Sample Work 1

Add   2.76
      + 3.75   ← Column form

Solution

```
  1 1
  2.76
+ 3.75   ← Column form
  6.51
```

Add from right to left.

Add 5 + 6 to get 11. Write 1 down and carry 1.

Next, add 7 + 7 to get 14 and add the 1 you carried to get 15.

Write down 5 and carry 1.

Finally, add 2 + 3 to get 5 and add the 1 you carried to get 6.

Sample Work 2

Add 0.837 + 0.213

Solution

0.837 + 0.213 ← Row form

```
  0.837
+ 0.213
───────
  1.050
```

Sample Work 3

Add 31.75 + 1.46 + 2.04

Solution

```
  31.75
   1.46
+ 2.04
───────
  35.25
```

Sample Work 4

Subtract 5.39 – 2.34

Solution

```
  5.39
− 2.34
──────
  3.05
```

Subtract 9-4 to get 5

Subtract 3-3 to get 0

Subtract 5-2 to get 3

Sample Work 5

Subtract — 6.39 + 61.12 ← Row form

Solution

$$
\begin{array}{r}
61.12 \\
-\ 6.39 \\
\hline
54.73
\end{array}
$$

Practice

The following problems are in row form. Arrange in column form and then add.

1. 45.2 + 2.396 + 150

2. 147.4 + 200 + 1.356 + 2.84

3. 2.3 + 2.1 + 87.26

4. 14 + 63 + 12.369

5. 16.269 + 12.53 + 0.36

6. 0.006 + 0.0003

7. 195.63 + 0.089 + 63.534

8. 95.0258 + 0.639

9. 0.0052 + 0.651

10. 2.09 + 3.91

11. 7.5 + 1.3

12. 1.79 + 4.14

13. 0.32 + 0.22 + 0.5

14. 0.25 + 6.5 + 2.7

15. 1.1 + 1.3

16. 5.5 + 9.6

17. 5.7 + 8.9 + 1.8

18. 2.5 + 3.4

19. 0.77 + 0.55 + 0.18

20. 6.7 + 9.2

21. Faico spent $2 for her uniform, $10 for a pair of shoes, and $1.30 for a pair of sacks. How much did she spend on all the items?

The following problems are in row form. Arrange in column form and then subtract.

1. 143.8 – 45.9

2. 53.18 – 5.6

3. 7.12 – 6.099

4. 66 – 32.0072

5. 98.25 – 52.914

6. 14 – 0009

7. 62.3 – 48.75

8. 65.3 – 41.76

9. Kebeh buys her uniform for $58.30. If she pays the businessman $100, how much change does she get?

10. Swen buys a bowl of rice for $1.50 and a piece of cassava for $2.20. How much change will he get back from $5?

11. A road construction company engineer is mapping out 2 sections of highway. The first is 55 miles long and the second is 38.9 miles long. How much longer is the first?

12. Gborze is 5 feet tall and Siryee is 4 feet tall. How much taller is Gborze than Siryee?

LESSON 3.2

MULTIPLYING AND DIVIDING DECIMALS

When multiplying decimal numbers, write the decimal point in the product according to the number of decimal places in the multiplier and multiplicand.

Sample Work 1

Multiply 3.241 x 0.23

Solution

3.241 x .23

```
  3.241   ← first factor has 3 decimal places
  X .23   ← second factor has 2 decimal places
  9723
 6482
 .74543  ← Decimal point is placed 5 places in the product since there are 5 decimal places in the
          multiplicand and multiplier.
```

Sample Work 2

Multiply 123 X 2.18

Solution

123 X 2 .18

```
    123   ← 0 decimal place
  X 2.18  ← 2 decimal places
    984
    123
    246
  268.14  ← 2 decimal places in the product.
```

We can do division of decimal just as we do division of whole numbers. Division problem can be written in three ways: using division sign, division bar, and division line. For example, 8 divided by 4 can be written in the following three ways:

1. $8 \div 4$,   2. $\dfrac{8}{4}$,   or $4\overline{)8}$

Sample Work 3

Divide $.84 \div 2$

Solution

$.84 \div 2$

```
        .42      ←——————— Quotient
    2 ).84       ←——————— Dividend
    |   8
    |  ——
        4
        4
      ——
        0        ←——————— remainder
```
Divisor

Sample Work 4

Divide $32.1 \div .121$

Solution

$32.1 \div .121$

$.121\overline{)32.1}$

Make the divisor whole number by moving the decimal point three places to the right in both the divisor and dividend. That is,

```
        265.29     ←——————— quotient
   121 )32100      ←——————— dividend
        242
        ———
        790
        726
        ———
        640
        605
        ———
         35         ←——————— remainder
```

Practice

Multiply

1. .03 X .03 X .03

2. .12 X .13

3. .04 X 0.4

4. .0273 X 100

5. 6.722 X 10

6. 3.228 X 2

7. 81.30 X 52.1

8. 5.123 X .14

9. 68.9 X 12

10. 352.1 X 24.1

Rewrite to make the divisor whole number.

11. $0.21\overline{)81}$ 12. $5.61\overline{)6.1234}$ 13. $1.3\overline{).146}$ 14. $.142\overline{)72}$

15. $3.22\overline{)85.4}$

divide

16. $4.3\overline{)4.601}$ 17. $.03\overline{).216}$

18. .076 ÷ 05

19. 9.281 ÷ .082

20. .7881 ÷ 3.7

21. 63 ÷ 6

CHANGING FRACTIONS TO DECIMALS

Fractions can be changed to decimals by using division. For example $\frac{3}{4}$ can be changed to decimal as follows:

$$\frac{3}{4} = 4\overline{)\begin{array}{l} .75 \\ 30 \end{array}}$$
$$\begin{array}{r} 28 \\ \hline 20 \\ 20 \\ \hline 0 \end{array}$$

so, $\frac{3}{4}$ is equivalent 0.75

Sample Work 1

Change to decimal $\frac{7}{20}$.

Solution: $\frac{7}{20}$

$$20\overline{)\begin{array}{l} 0.35 \\ 70 \end{array}}$$
$$\begin{array}{r} 60 \\ \hline 10 \\ 00 \\ \hline 100 \\ 100 \\ \hline 0 \end{array}$$

so, $\frac{7}{20} = 0.35$

Sample Work 2

Change to decimal $\frac{1}{16}$.

Solution: $\frac{1}{16}$ =

$$
\begin{array}{r}
.0625 \\
16\overline{\smash)10} \\
\underline{00} \\
100 \\
\underline{96} \\
40 \\
\underline{32} \\
80 \\
\underline{80} \\
0
\end{array}
$$

so, $\frac{1}{16}$ = .0625

Practice

Change the following fractions to decimals.

1. $\frac{5}{20}$

2. $\frac{2}{5}$

3. $\frac{1}{5}$

4. $\frac{7}{15}$

5. $\frac{6}{10}$

6. $\frac{3}{7}$

7. $\frac{3}{5}$

8. $\frac{1}{3}$

9. $\frac{3}{8}$

10. $\frac{5}{8}$

11. $\dfrac{1}{2}$

12. $\dfrac{1}{6}$

13. $\dfrac{1}{9}$

14. $\dfrac{1}{7}$

15. $\dfrac{1}{17}$

LESSON 3.4

ROUNDING OFF DECIMAL NUMBERS

Decimal numbers can be rounded off using the same steps for rounding whole numbers.

Steps:

1. First underline the digit in the place in which you want to round.

2. Mark or look at the digit one place to its right.

3. If the digit to its right is less than 5, round down.

4. If the digit to its right is 5 or more, round up.

Sample Work 1

Round to the nearest hundredth 0.6572

Solution

0.6572

0.6572 = 0.66

The number to the right of 5 is 7 (i. e. greater than 5), so round up.

Sample Work 2

Round to the nearest whole number 2.813

Solution

2.813

2.813 → 2.813 = 3

Practice

Round each number to the place of the underlined digit.

1. 3.<u>5</u>2

2. <u>5</u>.21

3. 0.<u>7</u>28

4. 3<u>0</u>.56

5. 15.0<u>9</u>8

6. 1<u>5</u>.67

7. 68.<u>0</u>6

8. 69.<u>2</u>5

9. 0.6<u>7</u>3

10. 6.<u>9</u>74

11. 2.<u>5</u>37

12. 4.<u>1</u>96

13. 12.3<u>8</u>3

14. <u>1</u>.58

15. 3<u>9</u>.518

LESSON 3.5

ESTIMATING DECIMALS AND WHOLE NUMBERS

We can estimate the answer of addition (sum) and subtraction (difference) when we do not need it to be exact. One way to do this is to round off each number first.

Sample Work 1

Round each addend to the nearest hundred and then add:

1312

684

069

Solution

1312 = 1300

684 = 700

096 = <u>100</u>

2,100

Sample Work 2

Round each addend to the nearest hundred and then add:

438 + 950

Solution

438 rounds down to 400

950 rounds up to 1000

400 + 1000 = 1,400.

Sample Work 3

Round each addend to the nearest dollar and then add:

$24.85 + $2.25 + $0.99

Solution

$24.85 = $25

$ 2.25 = $ 2.00

+$0.99 = $ 1

The total is $28.

Sample Work 4

Round each number to the nearest whole number and then subtract:

45.4 − 8.9

Solution

45.4 − 8.9

45 − 9 = 36

Practice

Round to the nearest hundred or hundredths and estimate each sum or difference.

1.   69.29 + 875.20

2.   6.312 + 8.253

3.   1.83 + 4.47

4.   3.85 + 12.25

5. $520 + $324 + $1378

6. 3291 + 243 + 881

7. 84.20 – 74. 50

8. 43.51 – 16.34

9. 5.86 – 3.19

10. 75.1 – 63.57

11. Round the minuend and subtrahend to the nearest hundred and then subtract:

    $85.60 – $64.30

UNIT 4

NUMBER THEORY

Learning points

In this unit, we shall

- learn how to square and find square roots of whole numbers;
- learn how to find the cube of whole numbers;
- learn how to find LCM and GCF; and
- learn about ratios and proportions.

LESSON 4.1

FINDING SQUARE AND PERFECT SQUARE ROOTS OF WHOLE NUMBERS

Squaring a number means to multiply that number by itself. For example, to square 3 means, 3 x 3. A calculator key can also be used to square a number.

Use the x^2 key on the calculator. If you want to square 4, first key in 4 and then press x^2 on the calculator.

Sample Work 1

Find the square of 5 using a calculator.

Solution

Key in 5 on the calculator and then press x^2. It will be 5 x 5 = 25.

Sample Work 2

Square the following numbers:

a.   6

b.   9

c.   10

Solution

a. $6 = 6 \times 6 = 6^2$

b. $9 = 9 \times 9 = 9^2$

c. $10 = 10 \times 10 = 10^2$

A perfect square is a number that is the square of a whole number. For example, 9 is a perfect square since $9 = 3 \times 3 = 3^2$.

A square root of a number is one of its two equal factors. For example, the square root of 16 is 4 because $16 = 4 \times 4$. The two equal factors for 16 are 4 and 4.

The symbol $\sqrt{}$ is used to show square root.

Sample Work 3

Find the perfect square root for 36.

Solution

$36 = \sqrt{36} = 6$.

Sample Work 4

Find the square root for 25.

The square root of 25 is 5 because $25 = 5 \times 5$

Sample Work 5

Find the following square roots:

a.   49

b.   1

c.   64

Solution

a. $49 = \sqrt{49} = 7$

b. $1 = \sqrt{1} = 1$

c. $64 = \sqrt{64} = 8$

Practice

Square the following numbers:

1. 7

2. 8

3. 11

4. 4.6

5. 12

6. 15

7. 2

8. 4

Find the square roots of the following numbers:

9. $\sqrt{81}$

10. $\sqrt{100}$

11. $\sqrt{121}$

12. $\sqrt{144}$

13. $\sqrt{400}$

LESSON 4.2

FINDING LCM AND GCF OF WHOLE NUMBERS

A multiple of a number is the product of that number and a whole number. To find the multiples of a number, multiply the number by 1, 2, 3, 4, and so on.

Sample Work 1

List the first 3 multiples of 2.

Solution

Multiply 2 by 1, 2, and 3. That is, 2 x 1 = 2, 2 x 2 = 4, 2 x 3 = 6.

The first 3 multiples of 2 are 2, 4, and 6.

A common multiple of numbers is a multiple of those numbers. The smallest number that is a multiple of two or more numbers is the least common multiple (LCM) of those numbers.

Sample Work 2

List the first 4 multiples of 3 and 4.

Solution

Multiply 3 and 4 each by 1, 2, 3, and 4.

Multiples of 3: 3 x 1 = 3; 3 x 2 = 6; 3 x 3 = 9; 3 x 4 = 12; 3, 6, 9, 12

Multiples of 4: 4 x 1 = 4; 4 x 2 = 8; 4 x 3 = 12; 4 x 4 = 16; 4, 8, 12, 16

The first 4 multiples of 3 are 3, 6, 9, 12.

The first 4 multiples of 4 are 4, 8, 12, 16.

The smallest multiple common to both numbers is 12 of 3 and 4.

The least common multiple of 3 and 4 is 12.

The greatest common factor (GCF) of numbers is the one number that evenly divides 2 or more numbers. You can find the GCF by first listing the prime factors of each number.

Sample work 3

Find the GCF of 6, 2, 8

Solution

Factors of 6: 1, 2, 3, 6

Factors of 2: 1, 2

Factors of 8: 1, 2, 4, 8

The largest factor common to 6, 2, and 8 is 2. So 2 is the GCF for 6, 2, and 8.

Practice

Find the least common multiples (LCM) of the following numbers.

1.   4 and 6

2.   8 and 4

3.   9 and 3

4.   5 and 10

5.   2 and 6

List the first three multiples of the following numbers.

6.   10 and 15

7.   15 and 3

8.   12 and 6

9.   4 and 10

10. 2, 3, and 4

11. 4, 6, and 3

12. 4, 5, and 20

13. 5, 15, and 3

Find the GCF for the following numbers.

14. 3, 5, 12

15. 3, 5, 6

16. 2, 4, 8

17. 30, 40, 20

18. 21, 3, 9

19. 15, 21, 9

LESSON 4.3

READING AND WRITING EXPONENTS

An exponent is used to show how many times a number is multiplied by itself.

For example, in 4^2, 2 is the exponent, and 4 is the base. The exponent is raised to the power of the base. It is written to the upper right hand corner of the base.

2^4 can be read as 2 to the 4th power.

Sample Work 1

Read the expression 3^2.

Solution

3^2 can be read as 3 to the 2nd power or 3 squared.

Sample Work 2

Write the base and the exponent: 5^3.

Solution

5^3

5 is the base, and 3 is the exponent.

Sample Work 3

Read and write the following:

a. 6^2

b. 2^3

Solution a. 6^2, read as 6 to the 2nd power or 6 squared.

Solution b. 2^3, read as 2 to the 3rd power or 2 cubed.

Practice

Read the following exponents:

1. 10^2

2. 6^3

3. 3^3

4. 3^4

5. 12^2

6. 9^2

Write the following exponents in words:

7. 5^4

8. 4^2

9. 13^2

10. 14^2

State or write the base and the exponent.

11. 4^4

12. 9^2

13. 5^4

14. 8^4

15. 20^2

16. 3^5

17. 2^5

LESSON 4.4

WRITING RATIOS AS FRACTION AND PERCENT

We use ratio to compare two numbers or items. For example, the ratio of men to women in a city is 5 to 15. Ratio can be expressed in fraction form, with a colon, or as a percent. Ratio is used in daily life to tell how big or small a quantity is in comparison with other quantities. Dieticians use ratio when following recipes. For example, a recipe for a seafood dinner calls for two cups of Argo oil to five pieces of crawfish.

Ratio is also used in business operations. For instance, Momo's pizza has 5 large slices and he eats 2. The ratio of part to the whole is 2:5.

Sample Work 1

Write the ratio in fraction form: 4 females to 5 males

Solution

4 females to 5 males = $\dfrac{4}{5}$

Sample Work 2

Write the ratio in colon form: 3 computer students to 4 building trade students.

Solution

3:4

Sample Work 4

Write the ratio as a percent: $\dfrac{1}{5}$

Solution

$$\frac{1}{5}$$

Divide 1 by 5

$1 \div 5 = 0.20$

= .20   move decimal 2 places to the right

20%   add percent sign

Practice

Write the ratios as fractions:

1.   1:2

2.   3:15

3.   5:10

Write the fractions as ratios in colon form

4. $\frac{1}{3}$

5. $\frac{2}{7}$

6. $\frac{3}{6}$

7. $\frac{10}{20}$

8. $\frac{30}{60}$

Write the ratios as percent.

9.   15 to 30

10.   3 to 9

11.   15 to 75

12. 80 to 100

13. 16 to 30

14. $\dfrac{1}{2}$

15. $\dfrac{1}{4}$

LESSON 4.5

WRITING AND SOLVING PROPORTION

When 2 ratios are equal, it is called a proportion. A proportion can be written in 2 ways, either as equal fractions ($\frac{1}{2} = \frac{5}{10}$) or by using a colon (1:2 = 5:10).

The following proportion is read as "3 is to 12 as 2 is to 8": $\frac{3}{12} = \frac{2}{8}$.

To solve problems involving proportions, use cross products to test whether the two given ratios are equal and whether they form a proportion. In other words, to know that two ratios are equal, find the cross products of the proportion. To find the cross product, cross multiply the outer terms, called the extremes, and the middle terms, called the means.

In the expression $\frac{3}{12} = \frac{2}{8}$, 3 and 8 are called the extremes, and 12 and 2 are called the means. Since the cross products of the extremes and the means are equal, it tells that the ratios are equal and that it is a true proportion.

$$\frac{3}{12} = \frac{2}{8}$$

The extremes and the means are indicated with arrows here.

Means
Extremes

Sample work 1

Tell whether the ratios are equal.

Solution

78

3 x 3 = 9 x 1 -- cross multiply

9 = 9

The two ratios are equal since their cross products are equal.

Sample Work 2

Tell whether the ratios are equal. $\dfrac{5}{20} = \dfrac{1}{4}$

Solution

5 x 4 = 20 x 1 --- cross multiply

20 = 20

The ratios are equal since the cross products are equal.

Sample work 3

In a shipment of 600 bags of rice, 10 are found to be inedible. How many inedible bags are expected in a shipment of 900?

Solution

$$\dfrac{600}{10} = \dfrac{900}{x}$$

600 (x) = 10 (900)

600x = 9000   — cross multiply

$$\dfrac{600x}{600} = \dfrac{9000}{600} \quad \text{— dividing each side by 600}$$

X = 15

15 inedible bags of rice are expected to be in the shipment.

Sample work 4

It takes Mr. Lasanah 30 minutes to design a test for his mathematics class. How long will he take to design tests for all five of his classes?

Solution

30 minutes to design a test

X minutes to design test for 5 classes

30 is to 1 as x is to 5

$$\frac{30}{1} = \frac{x}{5}$$

1 (x) = 30 (5)

X = 30 x 5

X = 150

It will take Mr. Lasanah 150 minutes to design tests for 5 classes.

Practice

1. If 5 pounds of chicken costs $5.20, how much will 15 pounds of chicken cost?

2. If a 10 cm long piece of cable wire weighs 60 grams, what will a 20 cm long of the same cable wire weigh?

3. If Mwangi can read 20 pages of a book in 25 minutes, how long would it take him to read 50 pages of the same book?

4. Sarah earns $2500 every 2 months working at an electronics store during the summer. What will she earn for five months working at the store?

5. The ratio of boys to girls in the class is 2 to 4. How many boys are there if there are 200 girls?

6. Dr. Kateh examined each of his patients for 30 minutes during an appointment last week. How many patients did he examine in 300 minutes?

Solve for x.

1. $\dfrac{x}{12} = \dfrac{2}{8}$ 2. $\dfrac{6}{10} = \dfrac{3}{x}$ 3. $\dfrac{x}{5} = \dfrac{12}{15}$

4. $\dfrac{3}{x} = \dfrac{4}{12}$ 5. $\dfrac{x}{3} = \dfrac{8}{12}$

Cross multiply, tell whether they are equal.

6. $\dfrac{2}{6} = \dfrac{3}{9}$ 7. $\dfrac{10}{15} = \dfrac{12}{18}$ 8. $\dfrac{6}{11} = \dfrac{4}{7}$

9. $\dfrac{3}{25} = \dfrac{1}{8}$ 10. $\dfrac{8}{15} = \dfrac{5}{9}$

Show the means and the extremes.

1. $\dfrac{3}{12} = \dfrac{2}{8}$ 2. $\dfrac{6}{10} = \dfrac{3}{5}$ 3. $\dfrac{x}{5} = \dfrac{12}{15}$

4. $\dfrac{3}{8} = \dfrac{4}{12}$

LESSON 4.6

DIVISIBILITY RULES FOR 2, 5, AND 10

A number is divisible by 2 if its last digit is 0, 2, 4, 6, or 8.

A number is divisible by 5 if its last digit is 0 or 5.

A number is divisible by 10 if its last digit is 0 or 10.

Sample Work 1

Is 236 divisible by 2?

Solution

Yes, 236 is divisible 2 since its last digit is 6.

Sample Work 2

Is 91 divisible by 2, 5, or 10?

Solution

No, 91 is not divisible by 2, 5, or 10 since its last digit is not an even number or 0.

Sample Work 3

Is 3,140 divisible by 2, 5, and 10?

Solution

Yes, 3,140 is divisible by 2, 5, and 10 since its last digit is 0.

Practice

Tell whether each number is divisible by 2, 5, or 10.

1. 84

2. 36

3. 125

4. 930

5. 286

6. 2,370

7. 19,576

8. 70,439

9. 26

10. 11,273

LESSON 4.7

FINDING CUBES OF WHOLE NUMBERS USING FACTORS METHOD

To find the cube root of a number, find a number that, when multiplied by itself three times, will give the original number. For example, to find the cube root of 27, find the number that when multiplied by itself three times will give 27. The cube root of 27, then, is 3, because $3 \times 3 \times 3 = 27$.

The cube of a number is the product of the number when multiplied by itself three times.

Sample Work 1

What is the cube root of 8?

Solution

The cube root of 8 = 2 since 2 x 2 x 2 = 8.

Sample Work 2

Find the cube root of 64.

Solution

The cube root of 64 is 4 since 4 x 4 x 4 = 64.

Practice

Find the cube root of the following numbers.

1. What is the cube root of 343?

2. What is the cube root of 216?

3.   What is the cube root of 125?

4.   What is the cube root of 512?

5.   What is the cube root of 729?

6.   What is the cube root of 1,000?

UNIT 5

MEASUREMENT

Learning points

In this unit, we shall

- learn how to change from one system of measure to another;
- learn how to add and subtract in English and Metric systems;
- learn how to find area, radius, and diameters of a circle; and
- learn how to find volume, perimeter, and circumference using formulas.

LESSON 5.1

READING AND WRITING ENGLISH AND METRIC SYSTEMS OF MEASURES

Units of measurement, such as inches, feet, yards, and miles, are used to measure length in the English system, while units such as meter, kilometer, centimeter, and millimeter are used for measuring length in the metric system.

Some relationships between units of lengths in the English system:

12 inches (in) = 1 foot (ft)

36 inches (in) = 1 yard (yd)

3 feet(ft) = 1 yard (yd)

5,280 feet (ft) = 1 mile (mi)

1,760 yards (yd) = 1 mile (mi)

Some relationships between units of mass in the English system:

16 ounces (oz) = 1 pound (1b)

2,000 pounds (1b) = 1 ton

The starting unit for length in the metric system is the meter. Meter is used when a smaller measuring unit is needed (such as when an inch is used in the English system).

Prefixes are added to meter to create different units in the metric system. Some of the units used for length in the metric system of measurement include:

meter (m) — an equivalent unit to yard or foot

centimeter (cm) — $\frac{1}{100}$ of a meter

millimeter (mm) — $\dfrac{1}{1000}$ of a meter

kilometer (km) — 1,000 meters. (It is used to measure very long distances the same way the mile is used in the English system.)

Some relationships in the metric length

1 kilometer (km) = 1,000 m

1 hectometer (hm) = 100m

1 decameter (dam) = 10m

10 decimeters (dm) = 1m

100 centimeters (cm) = 1m

1,000 millimeters (mm) = 1m

Some relationships between units of mass in the metric system

1 kilogram (kg) = 1,000g

1 hectogram (hg) = 100g

1 decagram (dag) = 10g

10 decigrams (dg) = 1g

100 centigrams (cg) = 1g

1,000 milligrams (mg) = 1g

1,000 kilogram (kg) = 1 metric ton

Practice

Write the abbreviations for the following units

1.   kilogram

2.   kilometer

3.   centimeter

4.   centigram

5.   decagram

6.   decameter

7.   hectometer

8.   hectogram

9.   milligram

10. millimeter

11. yard

12. pound

13. mile

14. ounces

15. feet

16. ton

LESSON 5.2

ADDING AND SUBTRACTING IN ENGLISH AND METRIC SYSTEMS

To add or subtract measures, one should know how to change from one unit to another between the units that are given. Add or subtract like units. To add or subtract, we need to remember unit conversions such:

10 decimeters (dm) = 1 meter

100 centimeters = 1 meter

60 seconds = 1 minute

60 minutes = 1 hour

When there are different units, convert the units, starting with the smallest unit, whether in the English system or Metric system.

Conversions in the Metric system are based on multiples of ten.

Sample Work 1

Add   2 hr 48 min 50 sec
           + 3 hr 20 min 25 sec

Solution

  2 hr 48 min 50 sec

+ 3 hr 20 min 25 sec

  5 hr 68 min 75 sec

= 5 hr 69 min 15 sec — adding 60 sec, which equals 1 min, to 68 min

= 6 hr 9 min 15 sec — adding 60 min, which equals 1 hr to 5 hr

Final answer = 6 hr 9 min 15sec

Sample Work 2

Subtract 10 hr 62 min 15 sec from_15 hr 125 min 98 sec

Solution

15 hr 125 min 98 sec

—10 hr 62 min 15 sec

5 hr 63 min 83 sec

5 hr 63 min 83 sec = 5 hr 64 min 23 sec — adding 60 sec to 63 min

= 6 hr 4 min 23 sec — adding 60 min to 5 hr

= 6 hr 4 min 23 sec

Sample Work 3

Add 3 m 11 cm + 5 m 21 cm

Solution

3 m 11 cm + 5 m 21 cm

Convert to have common unit — unit in cm

1 m = 100 cm; so, 3 m = 300 cm, 5 m = 500 cm

3 m 11 cm + 5 m 21 cm = 300 cm 11 cm + 500 cm 21 cm

= 311 cm + 521 cm

= 832 cm

Sample Work 4

Add 2 m 51 m 110 cm + 7 cm 8 m 38 cm

Solution

2 m 51 m 110 cm + 7 cm 8 m 38 cm

2 m 110 cm

51 m 7 cm

+ 8 m 38 cm
_____

61 m 155 cm

1 m = 100 cm; so, 155 cm = 1 m 55 cm

61 m 155 cm = 62 m 55 cm — adding 1 m to 61 m

Final answer = 62 m 55 cm

Sample Work 5

Subtract 7 km 215 dam

       — 4 km 115 dam
_____

Solution

   7 km 215 dam

— 4 km 115 dam
_____

   3 km 100 dam

100 dam = 1 km

3 km 100 dam = 3 km 1 km

              = 4 km

Practice

1. Add 2 hr 48 min 50 sec

   + 3 hr 20 min 25 sec
   _____

2. Add 5 hr 68 min 75 sec

   +2 hr 67 min 74 sec
   _____

3. Add 10 m 200 cm + 300 cm 10m

4. Add 10 m 150 m 120 cm + 78 m 200 cm

5. Add 5 m 400 cm + 15 m 100 cm

6. Subtract 9 km 205 dam

   —5 km 150 dam
   _____

7. Subtract 25 hr 145 min 95 sec

   —15 hr 67 min 15 sec
   _____

8. Subtract 7 cm 38 cm from 3 m 51 m 110 cm

LESSON 5.3

METRIC PREFIXES AND CONVERTING WITHIN METRIC SYSTEM OF MEASUREMENTS

Prefixes from 1,000 times larger to 1,000 times smaller:

Kilo is 1,000 times larger

Hecto is 100 times larger

Deca is 10 times larger

meter/gram

Deci is 10 times smaller

Centi is 100 times smaller

Milli is 1,000 times smaller

Notice that the prefixes show that from the unit (meter or gram) to the left-hand side are larger units and from the unit (meter or gram) to the right-hand side are smaller units. The base units used in the Metric system of measurement are meter and gram. Meter measures distance and gram measures mass.

Determine whether the initial unit is larger or smaller than the desired units (the unit you want to convert to) by looking at the line of prefixes and to decide whether it is to the right or left of the initial units (the unit you have). If it is to the right, you are converting from a larger to smaller unit. If it is to the left, you are converting from a smaller to larger unit.

For example, when computing how long 3 kilometers is in centimeters, look at the line of prefixes; you notice that "centi" is to the right of "kilo." Since the unit to convert to is to the right of the initial units, you are converting from the larger to smaller unit.

When converting between units in the Metric system, take the following steps:

1. Identify the unit you have.

2. Identify the unit you want to convert to.

3. Count the number of units between the unit you have and the unit you want to convert to.

4. Multiply successively by 10 if you are converting from a larger unit to a smaller one, and divide if you are converting from a smaller unit to a larger one.

For example, the distance between kilo and centi is 5 jumps. So, it is 10^5, which means 10 x 10 x 10 x 10 x 10.

$10^5 = $ 10 x 10 x 10 x 10 x 10 = 100,000

100,000 centimeters = 1 kilometer.

The table below summarizes the prefixes, suffixes, and units measuring distance and mass.

Measuring distance	Measuring mass	Prefixes
Kilometer	kilogram	kilo
hectometer	hectogram	hecto
decameter	decagram	deca
meter	gram	--unit--
decimeter	decigram	deci
centimeter	centigram	centi
millimeter	milligram	milli

Converting from one metric unit to another can be easily done by multiplying or dividing the unit you have by the appropriate power of 10 or multiples of 10. The number of spaces between the units of measurements tells the power of 10. For example, to convert centimeters into meters, divide the number of centimeters by 100 since there are 2 jumps between centimeter and meter. To convert meters into decimeters, multiply the number of meters by 10 since there is only 1 jump or space.

Sample work 1

Convert 5 kilometers to centimeters.

Solution

5 km to cm

There are 5 spaces from kilometer to centimeter (from left to right)

5×10^5 cm = 5 × 100,000 cm

= 500,000 cm

So, 5 km is equal to 500,000 cm.

Sample Work 2

Convert 8 hectometers to millimeters.

Solution

In this case, we are counting from the larger unit to the smaller unit.

We multiply the number of hectometers, 8, by 100,000 (i.e., 10^5)

8 hectometers to millimeters = 8×10^5

8 hectometers to millimeters = 8 x 10 x 10 x 10 x 10 x 10

8 hectometers to millimeters = 8 x 100,000

8 hectometers to millimeters = 800,000

8 hectometers = 800,000 millimeters

Sample Work 3

Convert 3 decimeters to decameters.

Solution

In this case, we count from the smaller unit to the larger unit.

We divide the number of decimeters, 3, by 100 (i.e., 10^2)

3 decimeters to decameters = $\dfrac{3}{100}$

3 decimeters to decameters = 0.03 decameter

3 decimeters = 0.03 decameter

Sample Work 4

Convert 5g to kg

Solution

There are three jumps to the left from gram to kilogram (i.e., from the smaller unit to the larger unit).

So you divide the number of grams (g), 5, by 1,000 (i.e., 10^3)

$5 \text{ g to kg} = \dfrac{5}{1000}$

So, 5 g = 0.005 kg

Practice

Arrange the following in order of size, starting with the smallest unit of measurement.

1.   7 g, 0.6 g, 100 g

2.   50 g, 1 g, 30 g

3.   6 kg, 5 kg, 4 kg

4.   2 mg, 1 mg, 2 mg

5.   4 mg, 10 mg

6.   3 g, 1 cg, 4 cg

7.   20 kg, 0.10 g, 5.1 kg

8.   75 cg, 4.9 cg, 100 cg

9.   7 mg, 0.5 mg, 30 mg

10.   0.9 kg, 0.10 kg, 0.5 kg

Arrange the following in order of size, starting with the largest unit of measurement.

11. 0.36 mm, 5 mm

12. 0.25 m, 5 mm, 20 mm

13. 6 m, .05 cm, 4 cm

14. 1 km, 2 km, .06 km

15. 35 m, .075 m, 3 m

16. 100 cm, 53.5 cm, 2 cm

17. 15 cm, 200 cm, 205 cm

18. 3 m, 5 m, 300 m

19. 20 km, 30 km, 100 km

20. 25 dam, 35 dam, 0.0300 dam

Convert the following

21. Convert 15 decimeters to decameters

22. Convert 3 decameters to millimeters

23. Convert 9 decigrams to kilograms

24. Convert 17 kilograms to centigrams

25. Convert 5 g to kg

26. Convert 10g to decag

27. Convert 20 cm to dam

28. Convert 50 cm to km

LESSON 5.4

FINDING AREAS OF TRIANGLES AND TRAPEZOIDS

To find the area of a triangle, multiply the base times half ($\frac{1}{2}$) of the height.

Area of triangle (A) = $\frac{1}{2}$ x base (b) x height (h)

A = $\frac{1}{2}$ x b x h

A = $\frac{1}{2}$ bh

Area is written in square units.

Sample Work 1

Find the area of the triangle with base 7 cm and height 5 cm.

Solution

A = $\frac{1}{2}$ x b x h

b = 7 cm, h = 5 cm

A = $\frac{1}{2}$ x 7 cm x 5 cm

A = $\frac{1}{2}$ x 35 cm^2

A = $\frac{35}{2}$ cm^2

A = 17.5 cm^2

Sample Work 2

Find the area of the triangle.

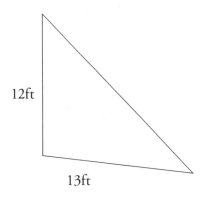

12ft

13ft

Solution

$A = \dfrac{1}{2} \times b \times h$

b = 13 ft, h = 12 ft

$A = \dfrac{1}{2} \times 13 \text{ ft} \times 12 \text{ ft}$

$A = \dfrac{1}{2} \times 156 \text{ ft}^2$

$A = \dfrac{156}{2} \text{ ft}^2$

$A = 78 \text{ ft}^2$

A trapezoid is a quadrilateral with one pair of opposite sides that are parallel.

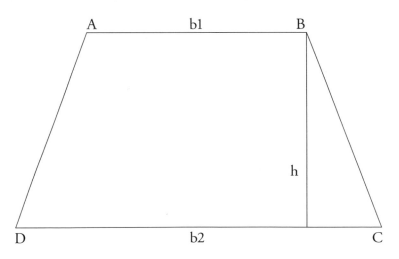

A trapezoid has four sides.

It has two opposite sides that are parallel.

In the diagram above, AB is parallel to DC.

The parallel sides are the bases of the trapezoid, b_1 and b_2.

The area of a trapezoid is equal to $\frac{1}{2}$ the height times the sum of the bases.

Area of Trapezoid (A) = $\frac{1}{2}$ x h $(b_1 + b_2)$

$A = \frac{1}{2}$ x h $(b_1 + b_2)$

Sample Work 3

Find the area of the trapezoid.

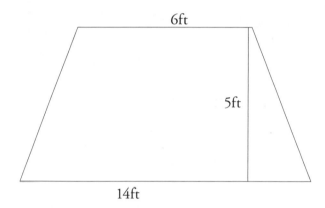

Solution

$A = \frac{1}{2}$ x h $(b_1 + b_2)$

$A = \frac{1}{2}$ x 5 ft (6 ft + 14 ft)

$A = \frac{1}{2}$ x 5 ft (20 ft)

$A = \frac{1}{2}$ x 100 ft^2

$= \frac{100}{2}$ ft^2

$= 50$ ft^2

The area is 50 ft^2.

Sample Work 4

Find the area: b_1 = 5 in, b_2 = 9 in, h = 15 in

Solution

$A = \dfrac{1}{2} \times h\ (b_1 + b_2)$

$A = \dfrac{1}{2} \times 15\ \text{in}\ (5\ \text{in} + 9\ \text{in})$

$A = \dfrac{1}{2} \times 15\text{in}\ (14\ \text{in})$

$A = \dfrac{1}{2} \times 210\ \text{in}^2$

$A = \dfrac{210}{2}\ \text{in}^2$

$A = 105\ \text{in}^2$

The area is 105 in^2.

Practice

Find the area of the triangles.

1. b = 15 m, h = 12 m

2. b = 8 ft, h = 20 ft

3. b = 5 in, h = 12 in

4. b = 6 yd, h = 4 yd

5. b = 20 m, h = 25 m

6. b = 13 ft, h = 17 ft

7. b = 10 in, h = 15 in

8. b = 2 yd, h = 5 yd

9.

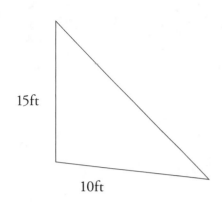

15ft

10ft

10.

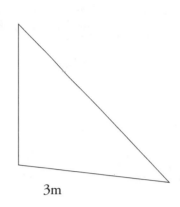

4m

3m

Find the areas of the trapezoid with the given lengths.

1. $b_1 = 7$ cm, $b_2 = 14$ cm; $h = 6$ cm

2. $b_1 = 9$ ft, $b_2 = 16$ ft ; $h = 18$ ft

3. $b_1 = 22$ in, $b_2 = 5$ in; $h = 3$ in

4. $b_1 = 24$ m, $b_2 = 9$ m; $h = 12$ m

5. $b_1 = 12$ cm, $b_2 = 5$ cm; $h = 8$ cm

6.

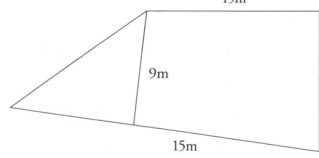

15m

9m

15m

7.

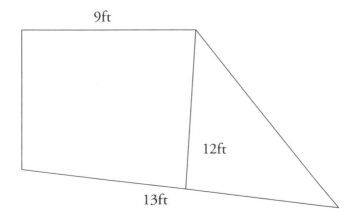

9ft

12ft

13ft

LESSON 5.5

COMPUTING DIMENSIONS OF A CIRCLE

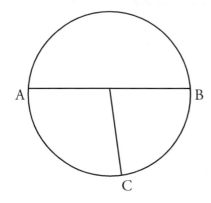

A circle is a closed curve with all points the same distance from the center. The diameter of a circle is the distance across the circle passing through the center.

In the circle above, line AB is the diameter, and from point C to line AB is the radius of the circle.

The area of a circle is equal to π times the radius squared.

Area of circle (A) = π x (radius)^2

A = π x r^2

A = πr^2

Sample Work 1

Find the area when the radius is 8 m.

Solution

A = π x r^2

r = 8 m

$A = 3.14 \times (8 \text{ m})^2$

$= 3.14 \times 8 \text{ m} \times 8 \text{ m}$

$= 3.14 \times 64 \text{ m}^2$

$= 200.96 \text{ m}^2$

The area is 200.96 m^2

Sample Work 2

The radius of a circle is equal to the diameter divided by 2.

$$\text{Radius (r)} = \frac{\text{diameter}}{2}$$

$$r = \frac{d}{2}$$

Find the radius of a circle with a diameter of 8 ft.

Solution

$$r = \frac{d}{2}$$

$$r = \frac{8}{2} \text{ ft}$$

$r = 4$ ft

The radius is 4 ft .

Sample Work 3

The diameter of a circle is equal to two times the radius.

Diameter(d) = 2 x radius (r)

$d = 2 \times r$

$d = 2r$

Find the diameter of a circle with a radius of 50 in.

Solution

d = 2 x r; r = 50 in

d = 2 x 50 in

d = 100 in

The diameter is 100 in.

The distance around the outside of a circle is called the circumference. Sometimes the circumference is called the perimeter of a circle. This distance can be calculated using the radius (r) formula, $C = 2\pi r$, where C is the circumference, r is the radius, and π is the ratio of the circumference to the diameter. The answer can be written using units of length such as yards, feet, inches, centimeters, or meters.

distance around the circle

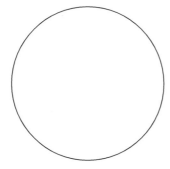

You can also use the formula $C = \pi$ x d to find the circumference of a circle.

$C = \pi$ x d; C = circumference

d = diameter of the circle

π = Greek letter (value: 3.14 or $\frac{22}{7}$)

Sample Work 4

Find the circumference if the diameter is 12 cm.

Solution

$C = \pi$ x d

d = 12 cm, π = 3.14

C = 3.14 x 12 cm

C = 37.68 cm

Sample Work 5

Find C, if d = 10 ft.

Solution

C = π x d

d = 10 ft, π = 3.14

C = π x d

C = 3.14 x 10ft

C = 31.4 ft

Practice

Find the area of each circle.

1. r = 6 cm

2. r = 5 ft

3. r = 10 m

4. r = 24 m

5. r = 30 in

6. r = 4 ft

7. Draw a circle and label the diameter, radius, and circumference.

Find the radius of each circle.

8. d = 8 ft

9. d = 3 yd

10. d = 12 ft

11. d = 24 cm

Find the diameter of each circle.

12. r = 20 in

13. r = 14 ft

14. r = 18 m

15. r = 22 ft

16. r = 30 cm

Find the circumference of each circle.

17. The diameter of a circle is 30 inches. Find the circumference. Use π = 3.14

18. Find the circumference of the circle with diameter 14 feet. Use π = 3.14

19. Find C if

    A.   d = 15 cm

    B.   d = 16 m

    C.   d = 6 ft

    D.   d = 8 ft

    (Use π = 3.14)

POLYGONS

A polygon is any geometric shape that has at least 3 sides and whose sides are all straight.

Types of polygons

A 3-sided polygon is called a triangle.

A 4-sided polygon is called a quadrilateral.

A 5-sided polygon is called a pentagon.

A 6-sided polygon is called a hexagon.

A 7-sided polygon is called a heptagon.

An 8-sided polygon is called an octagon.

A 9-sided polygon is called a nonagon.

A 10-sided polygon is called a decagon.

Interior and Exterior Angles of a Polygon

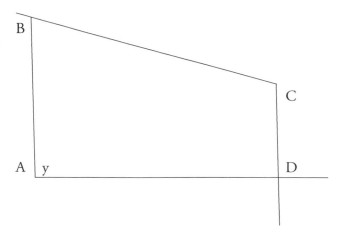

Interior Angle

The angles inside a polygon at the vertices are called interior angles of the polygon. They are the angles at each vertex on the inside of the polygon.

For example, in the figure on Page 111, angle y is an interior angle of quadrilateral ABCD.

Exterior Angle

Similarly in the figure, angle A is an exterior angle of quadrilateral ABCD.

The angles outside a polygon formed by the sides are called exterior angles of the polygon.

Interior Angles of Regular or Equilateral Polygons

A regular polygon is a polygon with equal sides and equal angles. The following are few common regular polygons (regular means all the sides and inside angles are equal).

A regular triangle is an equilateral polygon with all 3 sides and angles equal. Each angle measures 60 degrees.

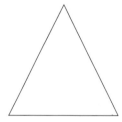

A regular quadrilateral is an equilateral polygon with all 4 sides and angles equal. Each interior angle measures 90 degrees. Draw it.

A regular pentagon is an equilateral polygon with all 5 sides and angles equal. Each interior angle measures 108 degrees.

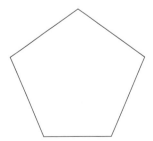

A regular hexagon is an equilateral polygon with all 6 sides and angles equal. Each interior angle measures 120 degrees. Draw it.

A regular heptagon is an equilateral polygon with all 7 sides and angles equal. Each interior angle measures 128.6 degrees.

A regular octagon is an equilateral polygon with all 8 sides and angles equal. Each interior angle measures 135 degrees.

A regular nonagon is an equilateral polygon with all 9 sides and angles equal. Each interior angle measures 140 degrees.

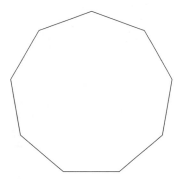

A regular decagon is an equilateral polygon with all 10 sides and angles equal. Each interior angle measures 144 degrees.

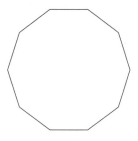

Types of Triangles, Common Polygons

Triangles: Three-sided polygons (triangles) with straight sides are the most basic shapes. They are classified on the basis of their sides and angles.

Equilateral triangle has 3 sides and 3 angles measuring the same length and angles.

Equilateral Triangle

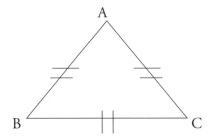

Angle A, angle B, and angle C are equal.

Side AB, side BC, and side CA are equal.

Isosceles triangle has 2 sides and 2 angles measuring the same length and angles.

Isosceles Triangle

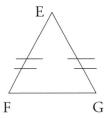

Angles F and G are equal.

Sides EF and GE are equal.

Scalene triangle has 3 sides and 3 angles measuring different lengths and angles.

Scalene Triangle

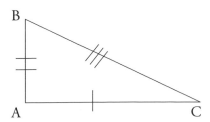

Right: A right triangle has 1 right angle. A right triangle may be scalene or isosceles.

Right Triangle

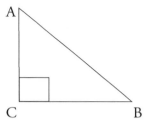

Quadrilateral is a geometric shape that has four straight sides. It is one of the most common geometric shapes. If you look around, you will notice that most rooms, doors, windows, tabletops, and other objects in our offices, houses, and apartments are quadrilaterals.

Some Common Quadrilaterals

Square is a geometric shape that has 4 right angles and 4 sides of equal measures. The sides opposite directly each other are parallel.

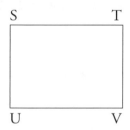

The 4 sides are equal

Sides SU and TV are parallel.

Sides ST and UV are parallel.

Sides SU, TV, ST. and UV are equal.

Angles S, U, T, and V are equal—90 degrees each.

Rectangle is a geometric shape that has 4 right angles. The 2 pairs of opposite sides are parallel, with adjacent sides having different lengths and measures.

Sides KL and MN are parallel and equal.

Sides KM and LN are parallel and equal.

Angles K, L, M, and N are equal.

Rhombus is a geometric shape that has 4 sides, all equal in length. The pairs of opposite sides are parallel. The opposite angles are equal.

Parallelogram is a geometric shape that has 4 sides. The pairs of opposite sides are equal in length, and they are parallel.

Parallelogram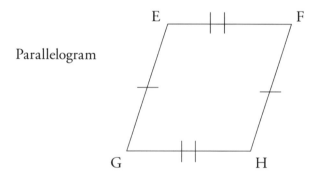

Sides EF and GH are parallel and equal.

Sides EG and FH are parallel and equal.

Trapezoid is a geometric shape that has at least 2 opposite sides that are parallel. In the figure below, sides NM and KL are parallel.

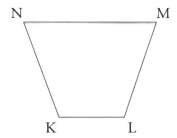

Practice

1. Draw a 3-sided polygon and give its name.

2. Draw a 4-sided polygon and give its name.

3. Draw a 5-sided polygon and give its name.

4. Draw a 6-sided polygon and give its name.

5. Draw a trapezoid and identify the sides parallel to each other.

6. Draw a rectangle and identify the sides parallel to each other.

7. Draw a rectangle and identify the sides equal to each other.

8. Draw a right triangle and show the right angle.

9. Draw an equilateral triangle and show the angles equal to each other.

10. Draw an equilateral triangle and show the sides equal to each other.

11. Draw an image of rhombus and label the opposite sides that are paralleled and show the equal angles.

12. Draw quadrilateral DEFG and show that angle E is an exterior angle.

UNIT 6
GEOMETRY

Learning Points

In this unit, we shall

- learn about the concept of space as a set of points;
- construct geometry figures; and
- learn about the usefulness of geometric figures in engineering and constructions.

LESSON 6.1

CONSTRUCTING AND IDENTIFYING LINES, RAYS, AND ANGLES

We can construct or draw lines in an attempt to measure objects. Carpenters, masons, tile layers, and building and road construction engineers use lines to measure and perform their works. They measure the lengths and widths of doors and windows of buildings using the concept of lines. Road construction engineers use lines to show demarcations between vehicle lanes that control the movement of traffic.

Intersecting lines meet or cross each other. For example, the figure below illustrates intersecting lines.

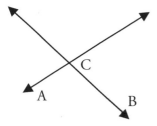

Lines A and B meet or cross each other at point C.

Parallel lines are two lines that do not meet no matter how far they extend. For example, lines M and N below illustrate parallel lines.

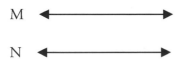

Lines M and N cannot meet no matter how far they extend in both directions.

Even if we construct three parallel lines like the ones below, they cannot still meet or intersect.

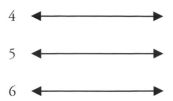

Lines 4, 5, and 6 are parallel lines. They will never meet no matter how far they extend.

Perpendicular lines meet at a right angle (90degrees). Lines E and F below illustrate perpendicular lines.

A **ray** is a line that has an end point and goes on forever in one direction.

Point B is the end point of this ray.

Lines are used to form angles such as vertical angles, opposite angles, interior angles, exterior angles, and so on.

Two rays meet to form an angle. The symbol < is used to represent an angle.

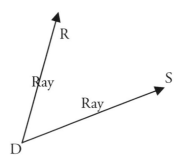

Rays R and S form an angle at D.

Vertical Angles

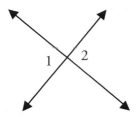

< 1 and < 2 are vertical angles. Vertical angles are equal. Angle 1 and angle 2 are equal (i.e., < 1 = < 2).

Vertical angles can also be referred to as opposite angles.

Use the figure below to describe the different angles.

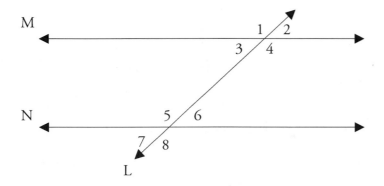

In the figure, the different angles are:

Interior Angles

$\angle 3$, $\angle 4$, $\angle 5$, and $\angle 6$ are **interior angles**

Alternate interior angles

$\angle 3$ and $\angle 6$

$\angle 4$ and $\angle 5$

Alternate interior angles are equal. So in the figure, $\angle 3 = \angle 6$;

$\angle 4 = \angle 5$.

Exterior Angles

$\angle 2$ and $\angle 8$

$\angle 1$ and $\angle 7$

Alternate exterior angles

∠2 and ∠7

∠1 and ∠8

Alternate exterior angles are equal. So in the figure, ∠2 = ∠7

∠1 = ∠8.

Pairs of corresponding angles

∠1 and ∠5

∠2 and ∠6

∠3 and ∠7

∠4 and ∠8

Corresponding angles are equal. So, in the figure, ∠1 = ∠5; ∠2 = ∠6; ∠3 = ∠7; ∠4 = ∠8.

Other Angles and Their Measures

Right angle measures 90 degrees.

Right angle:

Straight angle measures 180 degrees.

Straight angle:

Acute angle measures greater than zero degrees and fewer than 90 degrees.

Acute angle: <

Obtuse angle measures greater than 90 degrees and fewer than 180 degrees.

Obtuse angle:

Practice

1. Draw two intersecting lines meeting at P and R.

2. Draw four parallel lines.

3. Draw two perpendicular lines and show the right angle.

4. Draw two rays forming an angle at T.

5. Draw a straight angle.

6. Draw a vertical angle and show the vertical angles.

7. Draw and label interior angles.

8. Draw and label exterior angles.

9. In the figure below, identify and list the following angles, using lines L and M each:

   a. interior angles,

   b. alternate interior angles,

   c. exterior angles,

   d. alternate exterior angles,

   e. pairs of corresponding angles.

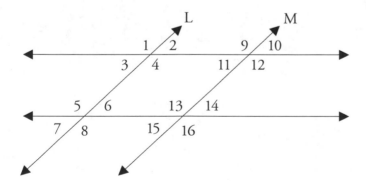

LESSON 6.2

FINDING PERIMETERS OF SQUARES, TRIANGLES, AND RECTANGLES

The perimeter of a square is the sum of the lengths of the 4 sides. The perimeter of a triangle is the sum of the lengths of the 3 sides.

Perimeter of triangle (p) = a + b + c

Perimeter of square (p) = 4 s or s + s + s + s

Perimeter of rectangle (p) = 2 l + 2 w

Sample Work 1

Find the perimeter of the square

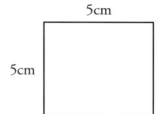

Solution

P = 4 s or s + s + s + s

P = 5 cm + 5 cm + 5 cm + 5 cm

P = 10 cm + 10 cm

P = 20 cm

The perimeter is 20 cm.

Sample Work 2

Find the perimeter of a square with each side measuring 6 feet.

Solution

P = 4 s

s = 6 ft

P = 4 x s

P = 4 x 6 ft

P = 24 ft

The perimeter of the square is 24 ft.

Sample Work 3

Find the perimeter of the triangle

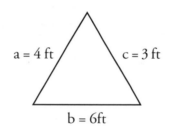

a = 4 ft   c = 3 ft

b = 6ft

Solution

Perimeter of triangle (P) = a + b + c

$\qquad$ = 4 ft + 6 ft + 3 ft

$\qquad$ = 4 ft + 9 ft

$\qquad$ = 13 ft

The perimeter is 13ft.

Sample Work 4

Find the perimeter of the triangle with each side measuring a = 7 in, b = 9 in, and c = 10 in.

Solution

Perimeter (p) = a + b + c

a = 7 in, b = 9 in, c = 10 in

P = 7 in + 9 in + 10 in

P = 7 in + 19 in

P = 26 in

The perimeter is 26 in.

Sample Work 5

Find the perimeter of the rectangle.

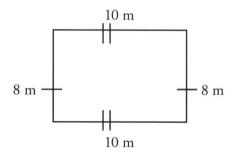

Solution

P = 2 l + 2 w, w = 8 m, l = 10 m

P = 2 x l + 2 x w

P = 2 x 10 m + 2 x 8m

P = 20 m + 16 m

P = 36 m

The perimeter is 36 m.

Sample Work 6

Find the perimeter of a rectangle with the length measured as 15 cm and width measured as 12 cm.

Solution P = 2 l + 2 w

l = 15 cm, w = 12 cm

P = 2 x 15 cm + 2 x 12 cm

P = 30 cm + 24 cm

P = 54 cm

The perimeter is 54 cm.

Sample Work 7

A rectangular bedroom door is 8 feet long and 4 feet wide. What is the perimeter?

Solution

P = 2 l + 2 w

l = 8 ft, w = 4 ft

P = 2 x 8 ft + 2 x 4 ft

P = 16 ft + 8 ft

P = 24 ft

The perimeter is 24 ft.

Practice

Find the perimeter of the squares

1.   10ft   2.   7cm

10ft   7cm

3.   If s = 4 ft

4.   If s = 20 m

5.   Each side of a square window is 4 ft. What is the perimeter?

6.   Each side of a square room is 15 ft. What is the perimeter?

7.   Kiama's bean farm is a square shape, with the side measuring 9 meters. Find the perimeter of the farm.

Find the perimeters of the triangles.

8.

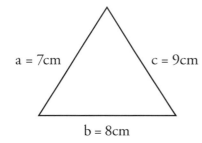

9.

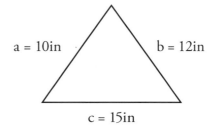

10.

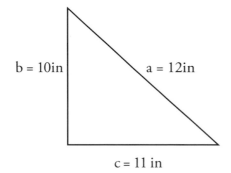

11. If a = 15 yd, b = 20 yd, c = 20 yd

12. If a = 14 m, b = 16 m, c = 18 m

13. The sides of a scalene triangle measure 25 cm, 26 cm, and 27 cm respectively. What is the perimeter?

14. The length of the base of an isosceles triangle measures 6 m and the length of each of the other 2 sides measure 8 m. What is the perimeter of the isosceles triangle?

15. Each side of an isosceles triangle has a length of 14.5 cm and its base has a length of 12.3 cm. Find the perimeter.

Find the perimeter of the rectangles:

16.   l = 12 ft

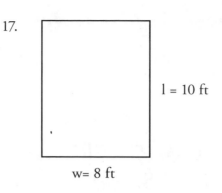

w =10 ft

17.

l = 10 ft

w= 8 ft

18.   If l = 20 ft, w = 10 ft

19.   If l = 9 cm, w = 7 cm

20.   If l = 30 m, w = 20 m

21.   If l = 15 in, w = 6 in

22.   A rectangular swimming pool has a length of 60 ft and a width of 50 ft. What is the perimeter?

23.   Seni's rectangular bedroom has a length of 20 ft and a width of 16 ft. What is the perimeter?

24.   The length of a rectangular window is 36 in. The width of the window is 24 in. What is the perimeter?

25.   A rectangular garden has a length of 100 cm and a width of 90 cm. What is the perimeter?

26.   A rectangular park has a length of 150 m and a width of 100 m. What is the perimeter?

LESSON 6.3

FINDING VOLUME OF PRISMS, CYLINDERS, AND SPHERES

We can find the volume of several solid geometric figures. The volume of a solid is the number of cubic units needed. Every 3-dimensional object will occupy certain space. The measure of such space is called its volume. The perpendicular distance between the bases is the height.

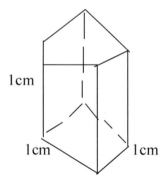

The volume of a prism is given by multiplying the area of the base times the height.

Formula: volume = area of base x height of prism

V = A x H

V = AH

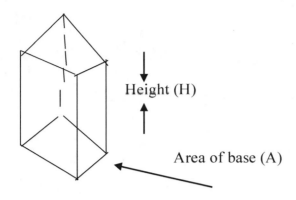

Height (H)

Area of base (A)

Sample Work 1

What is the volume of the rectangular solid whose base measures 2 ft by 3 ft and whose height is 4 ft?

Solution

Find the area of the rectangular base and multiply by the height.

Area of the rectangular base = length x width

A = l x w

A = 3 ft x 2 ft

A = 6 ft²

Volume = area x height

Volume = a x h

V = 6 ft² x 4 ft

V = 24 ft³

Sample Work 2

What is the volume of a prism whose bases are right triangles with legs measuring 3 in by 4 in and whose height is 8 inches?

Solution

Find the area of the triangular base and multiply by the height.

Area of the triangular base = $\frac{1}{2}$ x b x h

$A = \frac{1}{2}$ x 4 in x 3 in

$\frac{1}{2}$ x 12 in²

$\frac{12}{2}$ in²

6 in²

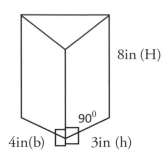

8in (H)

90⁰

4in(b)   3in (h)

Volume = AH

V = 6 in² x 8 in

V = 48 in³

The volume is 48 in³

The formula for finding the volume of a rectangular solid can also be written as V = lwh, where V = volume, l = length, w = width and h = height.

Sample Work 3

What is the volume of a rectangular solid 3 cm long, 2 cm wide, and 1 cm high?

Solution

V = lwh

V = 3 cm x 2 cm x 1 cm

V = 6 cm³

Sample Work 4

Find the volume of the triangular prism.

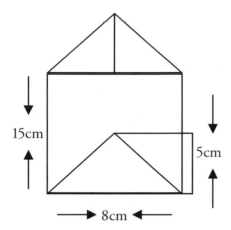

Solution

$V = \dfrac{1}{2} \times b \times h$

$V = \dfrac{1}{2} \times 8 \text{ cm} \times 5 \text{ cm} \times 15 \text{ cm}$

$V = \dfrac{1}{2} \times 600 \text{ cm}^3$

$V = 300 \text{ cm}^3$

The volume of the prism is 300 cm³.

Sample Work 5

A cylinder has a pair of parallel and circular bases with the same radii.

The formula for finding the volume of a cylinder is $V = \pi r^2 h$ or $V = \pi \times r^2 \times h$

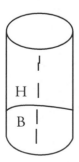

Area of base is πr^2

Find the volume of the cylinder.

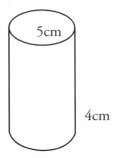

Solution

V = π r²h

V = 3.14 x (5 cm)² x 4 cm

V = 3.14 x 25 cm² x 4 cm

V = 3.14 x 100 cm³

V = 314 cm³

The volume is 314 cm³.

Sample Work 6

The formula for finding the volume of a cube is V = S³, where S = length of each side.

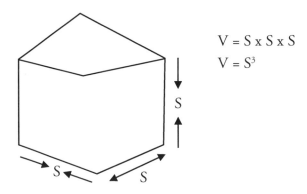

V = S x S x S
V = S³

What is the volume of a cube whose sides measure 3 cm each?

Solution

V = S³

$V = 3 \text{ cm}^3$

$V = 3 \text{ cm} \times 3 \text{ cm} \times 3 \text{ cm}$

$V = 3 \text{ cm} \times 9 \text{ cm}^2$

$V = 27 \text{ cm}^3$

Sample Work 7

The formula for finding the volume of a Sphere is, $V = \dfrac{4}{3} \times \pi \times r^3$ or $V = \dfrac{4}{3} \pi r^3$, $\pi = 3.14$

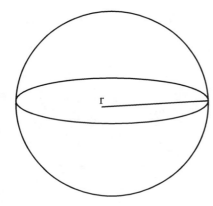

Find the volume of a Sphere whose radius is 4inches

Solution

$V = \dfrac{4}{3} \pi r^3$

$V = \dfrac{4}{3} \times 3.14 \, (4 \text{ in})^3$

$V = \dfrac{4}{3} \times 3.14 \times 64 \text{ in}^3$

$V = \dfrac{4}{3} \times 200.96 \text{ in}^3$

$V = \dfrac{803.84}{3} \text{in}^3$

$V = 267.9 \text{ in}^3$

Practice

Find the volume of the rectangular prisms. Use π = 3.14 and the formula V = lwh.

1. l = 3m, w = 10 m, h = 12 m

2. l = 20 cm, w = 9 cm, h = 30 cm

3. l = 24 cm, w = 12 cm, h = 13 cm

4. l = 6 in, w = 3 in, h = 2 in

Find the volume of the cylinders. Use the formula $V = \frac{4}{3} \times \pi \times r^3$, where π = 3.14

5. r = 10 yd, h = 15 yd

6. r = 13 cm, h = 8 cm

7. r = 4 ft, h = 3 ft

8. r = 2.3 cm, h = 3.2 cm

9. r = 3 m, h = 5 m

10. r = 4 m, h = 10 m

11. r = 6 in, h = 9 in

Find the volume of the spheres. Use π = 3.14.

12.

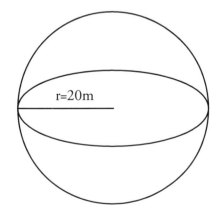

r=20m

13. r = 15 ft

14. r = 20 cm

15. r = 24 ft

16. r = 18 cm

17. r = 32 m

18. r = 6 in

USE OF GEOMETRIC FIGURES IN ROAD AND BUILDING CONSTRUCTIONS

Many geometric figures are used for controling and protecting the movement of vehicles and pedestrians. Few of those geometric figures are discussed in this section.

Side road intersection sign

We can relate this traffic sign to square or quadrilateral, which has four equal sides. This traffic sign shows that there is an intersecting road joining the road you are on so; the vehicle operator should slow down and take precaution.

Speed limit sign

We can relate this traffic sign to a rectangle, which has four sides with two opposite sides parallel and equal. Speed limit reflects the minimum or maximum mileage permitted in miles per hour (mph) or kilometers per hour (km/h)/or a vehicle operator should use when driving. We can also trace this traffic sign to the designs of rooms, doors and windows measurements of buildings, dining table top, etc. Sometimes soccer and football fields are measured in similar fashion like the speed limit sign design.

Railroad crossing sign

We can relate this traffic sign to circle. Railroad crossing signs reflect that you are approaching railroad tracks, so you must look and listen for trains in both directions before crossing the track. We can see practical use of circle on soccer field, 'centric circle', bicycle tire wheel ring, and roundabout. Car tires and wheels are also forms of circles.

Pedestrian crossing signs

We can relate this traffic sign to a square or quadrilateral, which has four equal sides. A pedestrian crossing sign is used for foot traffic. It is intended to maintain order on the streets and to protect drivers and pedestrians.

Road Traffic Sign images via courtesy of www.pixabay.com. Public Domain.

Practice

1. Draw a Right curve sign. We can relate this traffic sign to square, which has four equal sides.

   This traffic sign reflects that there is a right curve ahead and so, the vehicle operator should slow down and take precaution.

2. Draw a Crossroad sign. The crossroad sign reflects that another road crosses the road you are on, so you must look left and right for other traffic or pedestrians. We can relate this sign to a square or quadrilateral, which has four equal sides.

3. Draw three different road traffic signs you know and relate or trace each to geometric figure.

   Explain how these road signs (geometric symbols) are used to control or protect movement of human and vehicles.

UNIT 7

PROBABILITIES AND STATISTICS

Learning Points

In this unit, we shall

- learn how to find mean and discuss its usefulness in real life;
- learn how to find median and discuss its usefulness in real life;
- learn how to find mode and discuss its usefulness in real life; and
- learn about statistical graphs and probabilities.

FINDING MEAN

A mean is the average of a group of numbers. This group of numbers is also called a data set. We can use mean to find the average of student grades obtained during marking periods in school.

To find the mean or average, add all the numbers or grades and divide the sum by the number of grades.

M = sum of data divided by number of data

$$M = \frac{\text{sum of data}}{\text{number of data}}$$

The mean is useful because it is used to represent the typical value and can serve as a yardstick for making sense out of a situation. For example, if a school wants to know how many weeks, on average, a student attends school in a year, the school can find the mean of the weeks attendance of a group of students.

Sample Work 1

Find the mean for the numbers 72, 77, 88, 95

Solution

$$M = \frac{\text{sum of data}}{\text{number of data}}$$

$$M = \frac{72+77+88+95}{4}$$

$$= \frac{332}{4}$$

$$= 83$$

The mean for this student's grade is 83.

Sample Work 2

The second marking period test grades for Lavala is 80 in science, 79 in English, 68 in history, 90 in math, and 85 in trade. Find the mean for Lavala's grades.

Solution

Science	80
English	79
History	68
Math	90
Trade	<u>85</u>
Total (sum)	402

Total divided by number of subjects (5)

$$= \frac{402}{5}$$

$$= 80.4$$

Sample Work 3

The chart below shows the populations of five counties in the years indicated.

County	Year & Population		
	2017	2018	2019
Maryland	4,600	4,800	5,600
Sinoe	5,600	5,700	6,500
Grand Gedeh	1,300	1,620	2,000
Lofa	5,700	6,800	7,000
Bong	7,100	7,300	7,500

Use the data on the chart to answer the following questions:

A.   What is the mean of the populations in 2017?

B.   What is the mean of Maryland's population in those three years?

Solution A

Population in 2017 for the five counties: 4,600, 5,600, 1,300, 5,700, 7,100

$$\text{Mean 2017} = \frac{4600 + 5600 + 1300 + 5700 + 7100}{5}$$

$$\text{Mean 2017} = \frac{24300}{5}$$

Mean 2017 = 4,860

Solution B

$$\text{Mean of Maryland population} = \frac{4600 + 4800 + 5600}{3}$$

$$\text{Mean of Maryland population} = \frac{15000}{3}$$

Mean of Maryland population = 5,000

Practice

Find the mean.

1.   64, 56, 83, 90

2.   65, 82, 94, 94, 82

3. $8.25, $5.50, $7.35, $5.00, $20.98

4.   215, 220, 230, 235

5.   103, 123, 79, 89

6.   23, 16, 43, 25, 19

There are 4 different basketball teams and they play 3 games each. The scores of each team are recorded on the chart below. Use the chart to answer questions 7–11.

	Game 1	Game 2	Game 3
Team A	46	78	56
Team B	65	57	95
Team C	31	67	72
Team D	55	80	87

7. What is the mean of team B?

8. What is the mean of team C?

9. What is the mean of game 2?

10. What is the mean of game 3?

11. Liamon wants to join the team that is doing the best among the four teams. If the teams are ranked by their individual mean scores, which team would Liamon prefer to join?

The chart below shows the population of six cities in the years as indicated.

City	Year & Population		
	2016	2017	2018
Pierce	5,700	5,900	6,000
Yakima	6,500	6,700	6,900
Gray	2,000	3,020	4,000
Whatcom	7,500	8,300	8,000
Benton	8,300	8,700	8,900
King	9,000	9,700	9,800

Use the data on the chart to answer the following questions:

12. Find the mean of the population in 2016.

13. Find the mean of the population in 2017.

14. Find the mean of the population in 2018.

15. What is the mean of Pierce city's population in those three years?

16. What is the mean of Whatcom city's population in those three years?

17. Find the mean of Benton city's population in those three years?

18. Find the mean of King city's population in those three years?

LESSON 7.2

FINDING MEDIAN

Median is the middle number in a group of numbers when that group of number is arranged from small to big or from big to small.

Sample Work 1

Arrange the following set of number from smallest to greatest: 1, 3, 2, 5, 6, 9

Solution

Numbers arranged from smallest to greatest:

1, 2, 3, 5, 6, 9

Sample Work 2

Order the following numbers from greatest to smallest: 2, 4, 3, 1, 7, 8

Solution

Numbers arranged from greatest to smallest:

8, 7, 4, 3, 2, 1

Sample Work 3

Order the following population from greatest to least:

12,000, 30,000, 16,000, 10,000, 17,000, 19,000.

Solution

30,000, 19,000, 17,000, 16,000, 12,000, 10,000

Sample Work 4

Find the median of the numbers 1, 5, 6, 4, 3

Solution

Numbers from smallest to biggest

1, 3, 4, 5, 6

The middle number (median) is 4.

Sample Work 5

The first marking period grades for 7 students in the 2nd grade class are 85, 76, 81, 58, 69, 92, and 90. Find the median.

Solution

Grades ordered from small to big: 58, 69, 76, 81, 85, 90, 92

58, 69, 76, 81 , 85, 90, 92

81 is the middle number.

So, the median is 81.

Sample Work 6

When there are 2 middle numbers, find the median by adding the numbers and dividing by 2.

Find the median of the numbers 11, 17, 5, 9, 8, 12.

Solution

11, 17, 5, 9, 8, 12.

Arrange the numbers from big to small: 17, 12, 11, 9, 8, 5

The two middle numbers are 11 and 9. Add the two middle numbers and divide by 2.

$$\text{median} = \frac{11+9}{2}$$

The median is 10.

Sample Work 7

There are 3 different basketball teams, and each plays 5 games during the tournament. The score keepers record their scores as illustrated on the table below.

Team	Game 1	Game 2	Game 3	Game 4	Game 5
Lone Star	87	82	56	46	78
Black Star	80	75	95	80	80
Eagle	68	67	79	55	95

Use the information on the table to answer the following questions.

A.   What is the median score for the Lone Star team?

B.   What is the median score for the Eagle team?

Solution A

Lone Star scores: 87, 82, 56, 46, 78

Scores arranged from greatest to least: 87, 82, 78, 56, 46

Median = 78.

Solution B

Eagle scores: 68, 67, 79, 55, 95

Scores arranged from least to greatest: 55, 67, 68, 79, 95

Median = 68.

Practice

Find the median:

1. 6, 4, 7, 8, 2

2. 8, 9, 3, 10, 4

3. 8, 7, 10, 12, 14, 15

4. 6, 8, 9, 10, 11, 12

5. 20, 30, 32, 14, 25

6. 15, 12, 11, 10, 14, 7

7. The test scores for 5 students are 80, 72, 75, 90, and 60. Find the median score.

8. The ages of 7 students in the 3rd-grade class are 9, 10, 8, 7, 6, 11, and 8. Find the median age of the students.

9. Find the median score for 68, 78, 94, 85, 83, and 80.

10. The populations of 7 communities are recorded as 10,000, 20,000, 15,000, 11,000, 13,000, 9,000, and 14,000. Order the population from least to greatest.

11. Arrange the populations in question 10 from greatest to least.

12. Find the median of the populations in question 10.

13. There are 6 football teams, and each will play 5 games during the upcoming tournament. The score for each team will be recorded as shown on the table below.

Team	Game 1	Game 2	Game 3	Game 4	Game 5
Fargo Star	25	12	10	22	36
Kent Tiger	17	26	18	42	38
Zao Eagle	20	33	34	40	41
Eleven-Brother	15	21	27	35	40
Claw	35	42	40	37	28
Wolf	41	19	35	32	30

Use the information on the table to answer the following questions.

a.   What is the median score for the Fargo Star team?

b.   What is the median score for the Eleven-Brother team?

c.   What is the median score for the Kent Tiger team?

d.   What is the median score for the Claw team?

e.   What is the median score for the Wolf team?

f.   Find the median score for Game 3.

g.   Find the median score for Game 4.

h.   Find the median score for Game 5.

LESSON 7.3

FINDING MODE

The number or item that appears more than the other numbers in a group of number is called the mode. These numbers are called a data set. For example, the number 3 appears more than the other numbers in the following listing: 7, 4, 8, 3, 3, 3, 3, 6, 2. Notice that 3 occurs 4 times in the listing. So 3 is the mode.

A teacher would find the mode of her students' math test scores by listing all the scores in order from least to greatest or from greatest to least, and then finding the number that occurs on the list most often.

The mode has usefulness in areas such as printing. For example, the United Methodist School System might decide to print more copies of the most popular textbooks. Printing more equal numbers of copies for different textbooks could cause a shortage or oversupply of certain books.

Sample Work 1

Find the mode for the following numbers: 8, 7, 7, 6, 9, 7

Solution

The numbers are arranged from least to greatest:

6, 7, 7, 7, 8, 9

Seven appears more often than the other numbers, so 7 is the mode.

Sample Work 2

Find the mode of the grades 75, 83, 92, 65, 83, 71, 93.

Solution

The numbers are arranged from least to greatest:

65, 71, 75, 83, 83, 92, 93

Since 83 appears more than the other numbers, 83 is the mode.

Sample Work 3

The table below shows how many times 4 boys watched videos during the last week in December.

	Monday	Tuesday	Wednesday	Thursday	Friday
Sayku	1	1	2	2	1
Kroma	2	2	2	3	2
Bankalee	1	1	1	2	1
Sirleaf	3	5	2	3	1

A. Find the mode for Bankalee.

B. Find the mode for Wednesday.

Solution A

Bankalee's views for the week are 1, 1, 1, 2, 1

Arranging from least to greatest: 1, 1, 1, 1, 2

The number 1 appears most often than all the other numbers, so the mode is 1.

Solution B

Wednesday's views are recorded as 2, 2, 1, 2.

Arranging from greatest to least: 2, 2, 2, 1.

The number 2 appears most often than all the other numbers, so the mode is 2.

Practice

Find the mode for the following sets of data.

1. 2, 5, 2

2. 6, 4, 6, 3

3. 1, 1, 1, 7, 5, 6, 1

4. 9, 10, 12, 6, 10

5. 3, 2, 3, 5, 3, 1

6. 8, 6, 4, 6, 7, 3, 6

7. 1, 2, 3, 4, 2, 5, 6, 8

The table shows how many times 5 friends visited the movie theater each month.

	January	February	March	April	May
Charlie	1	1	2	2	2
Jamie	2	2	1	3	2
Dave	4	5	1	2	2
Tom	3	3	4	3	1
Sarah	5	1	3	1	2

Use the table to answer the following questions.

8. Find the mode for January.

9. Find the mode for March.

10. Find the mode for May.

11. Find the mode for Jamie.

12. Find the mode for Tom.

13. Find the mode for Sarah.

LESSON 7.4

DRAWING AND INTERPRETING BAR GRAPHS

A bar graph is used to show amounts. It is also used to show the number of students scoring certain grades.

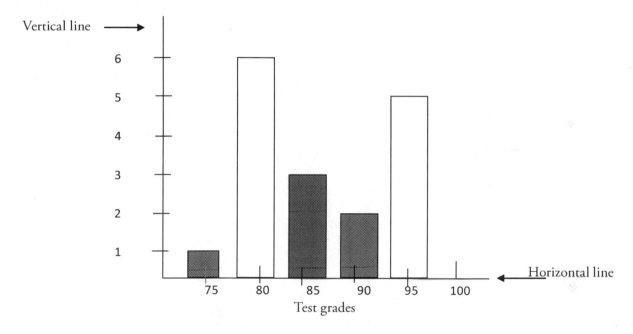

To draw a bar graph, draw horizontal and vertical lines meeting at right angle.

The vertical line shows the number of students, while the horizontal line shows the test grades. The height of each bar shows how many students obtained a particular grade.

Sample Work 1

Answer the following questions using the bar graph above.

1.   How many students scored 75?

2.   How many students scored 80?

3.   How many students scored 95?

157

Solution 1

Look at the bar for 75.

Find the height for the bar for 75.

One student scored 75.

Solution 2

Look at the bar for 80.

The height for the bar for 80 is 6.

Six students scored 80 each.

Solution 3

Look at the bar for 95.

The height for the bar for 95 is 5.

Five students scored 95.

Practice

I. Use the bar graph shown below to answer Questions 1-8

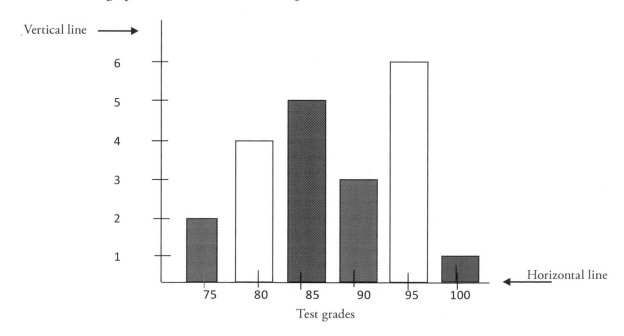

1.   How many students scored 85?

2.   How many students scored 75?

3.   How many students scored 95?

4.   How many students scored 90?

5.   How many students scored 100?

6.   How many students scored 80?

7.   How many students scored 93?

8.   How many students scored 76?

PROBABILITY

Probability is a measure that quantifies the chance of an event occurring. Probability is a guide that does not tell exactly what is going to happen. Some events cannot be predicted with total certainty. For example, what are the chances of rolling a 5 on a die? There is only 1 face with a 5 on the die, and there are 6 faces altogether, which is the total number of outcomes. So the probability $= \frac{1}{6}$.

Playing cards are also involved in probability. For example, the probability of picking an ace from a deck of cards is $\frac{4}{52}$ since a deck contains 52 cards, and there are 4 aces.

Sample Work 1

James rolls a 6-sided die once. What is the probability of rolling a number less than 2?

Solution

The probability of rolling a number less than 2 (i. e. p(less than 2)

The die has 6 sides, with the numbers 1, 2, 3, 4, 5, and 6 on it. The only number less than 2 is 1. There is 1 number less than 2.

$$P(\text{less than 2}) = \frac{1}{6}$$

Sample Work 2

You pick a card at random from a deck. What is the probability that you pick a king?

Solution

A deck contains 52 cards, and there are four kings.

$$P(\text{king}) = \frac{4}{52}$$

Sample Work 3

Peter has 7 marbles in his bag. Five are green and 2 are blue. What is the probability that he picks a green marble from the bag?

Solution

The number of times a green marble can be picked from the bag is 5 because there are 5 green marbles in total.

The total number of outcomes is 7 because there are 7 marbles altogether in the bag.

So the probability of picking a green marble = $\dfrac{5}{7}$.

Sample Work 4

What is the probability of landing on a number greater than 3 on a die?

Solution

The events (numbers) greater than 3 are 4, 5, and 6 on a die. There are 3 numbers greater than 3 on a die.

P (greater than 3) = $\dfrac{3}{6}$

Practice

1. Youmalay has 5 marbles in the bag. Three are red and 2 are green. What is the probability that Youmalay picks a green marble from the bag?

2. Massaquoi has 8 marbles in his bag. Seven are green and 1 is purple. What is the probability that he picks a purple marble from the bag?

3. Nancy has 4 marbles in her bag. Two are red and 2 are yellow. What is the probability that she picks a yellow marble from the bag?

4. What is the probability of landing on a number greater than 2 on a die?

5. What is the probability of landing on a number less than 4 on a die?

6. What is the probability of landing on a number greater than 6 on a die?

7.   What is the probability of landing on a number greater than 5 on a die?

8.   What is the probability of landing on a number less than 6 on a die?

9.   Mom rolls a 6-sided die once. What is the probability of rolling a number less than 1?

10. Bosco rolls a 6-sided die once. What is the probability of rolling a number less than 3?

11. Kojo rolls a 6-sided die once. What is the probability of rolling a number greater than 1?

12. Garmoyu picks a card at random from a deck. What is the probability that he picks a heart?

13. Frey picks a card at random from a deck. What is the probability that the card picked is a diamond?

Printed in the United States
By Bookmasters